ADVANCES IN ASTRONOMY AND ASTROPHYSICS

AN INTRODUCTION TO MOLECULAR CLOUDS

ADVANCES IN ASTRONOMY AND ASTROPHYSICS

Additional books and e-books in this series can be found on Nova's website under the Series tab.

ADVANCES IN ASTRONOMY AND ASTROPHYSICS

AN INTRODUCTION TO MOLECULAR CLOUDS

SACHIN KAOTHEKAR
EDITOR

NOTICE TO THE READER

Library of Congress Cataloging-in-Publication Data

ISBN: 978-1-53619-178-3

Published by Nova Science Publishers, Inc. † New York

CONTENTS

PREFACE

An Introduction to Molecular Clouds describes the formation of molecular clouds and the innovative features of molecular clouds with different physical parameters. In this book, Jean-gravitational instability is discussed with different physical parameters, which is the major cause of the formation of molecular clouds in the interstellar medium (ISM), and the way molecular clouds are formed in the astrophysical plasma environment is described. The authors aim to determine the basic conditions responsible for the formation of heavenly bodies in the universe. The book deals with radiative instability in a variety of conditions incorporating different physical parameters such as viscosity, rotation, permeability, porosity, thermal conductivity, Hall current, Finite ion Larmor radius corrections, finite electrical resistivity, radiative heat-loss functions and finite electron inertia, both in gaseous plasma and quantum plasma environments.

Chapter 1 - Dark globules have an important role in the evolution of the stars with low and/or middle masses. Many globules are connected with such stars, mainly with pre-main-sequence young stars (T Tauri type, Herbig A_e/B_e type et cet.). The investigation of rotation of the globules is important for clarifying the evolution of stars: as a rule, the specific angular momentum is much larger for the globules than for the binary and single stars. In this chapter the authors present the results of $^{12}CO(1\text{-}0)$ observations of several rotating dark globules. Angular velocities of investigated globules are in the range $(4 - 4.6)\cdot 10^{-14}$ s^{-1}, which is larger than the results obtained for the globules by other authors.

Chapter 2 - A new simplistic model equation of macroscopic equilibrium state for understanding the realistic gyro-equilibrium properties of degenerate neutron stars is semi-analytically developed. The adopted model setup prudently includes all the possible realistic active tonality agencies unavoidable in the course of realistic stellar fluid dynamics concurrently. As such, it conjointly includes thermal pressure, quantum degeneracy pressure, turbulence pressure, self-gravity, magnetic field, viscosity, and the Coriolis rotational effects on the relevant scales. A systematic combination of the fundamental and derived multi-parametric analyses renders the obtained resultant equation of macroscopic state in a standard normalized form of an ordinary differential (first-order) equation on the material density alone. It is numerically solved, emphatically analyzed and graphically illustrated. A direct non-trivial correlation of the density on the explicit dependencies on temperature, viscosity and electric current is portrayably established. The expounded semi-analytic findings match fairly with the existing non-homologous predictions extensively available in the literature.

Chapter 3 - A new barotropic model equation of the macroscopic equilibrium state for understanding the gyro-equilibrium properties of spherically symmetric stellar structures is semi-analytically developed. The adopted model setup judiciously includes all the possible realistic stellar parametric factors concurrently, such as non-local self-gravity, fluid turbulence, viscosity, Coriolis rotational effects and relevant radiative hydrodynamical effects. The obtained resultant radial differential equation of state in a standard normalized form is numerically solved as an initial value problem and portraylly illustrated. Despite the accounted complications, a direct correlation of the net pressure composed of both the material and radiative hydrodynamic contributions is analyzed and established. The expounded semi-analytic findings fairly match with the existing predictions available in the literature as special astro-corollaries. The proposed analysis can be useful to study various thermophysical features of the stars and their ambient atmospheres.

Chapter 4 - The effect of rotation, porosity and finite ion Larmor radius (FLR) corrections on the gravitational instability and radiative instability of infinite homogeneous plasma has been explored integrating the consequences of radiative heat-loss function and thermal conductivity. The general dispersion relation is obtained by means of the normal mode analysis scheme with the help of appropriate linearized perturbation equations of the predicament. This dispersion relations is further condenses for rotation axis parallel and perpendicular to the magnetic field. Stability of the medium is argued by pertaining Routh Hurwitz's criterion and it is found that Jeans criterion establishes the stability of the medium. The authors locate that the presence of radiative heat-loss function and thermal conductivity amend the fundamental Jeans criterion of gravitational instability into radiative instability criterion. Numerical computations have been executed to show the effect of various parameters on the growth rate of the Jeans-gravitational instability. The authors find that rotation, FLR corrections and medium porosity steady the growth rate of the organization in the transverse mode of propagation. The authors' result demonstrates that the rotation, porosity and FLR corrections affect the dens molecular clouds configuration and star formation.

Chapter 5 - The problem of thermal instability is investigated for a partially ionized plasma which has connection in astrophysical condensations for the formation of molecular clouds in interstellar medium (ISM). The normal mode analysis technique is used in this problem. The general dispersion relation is obtained from linearized perturbation equations of the problem. Effects of collisions with neutrals, radiative heat-loss function, viscosity, thermal conductivity and magnetic field strength, on the thermal instability of the system are discussed. The conditions of instability are obtained for heat-loss function with thermal conductivity. Numerical computations have been performed to calculate the effect of different physical parameters on the growth rate of the thermal instability of considered system. The heat-loss function, thermal conductivity, viscosity, magnetic field and neutral collision have stabilizing effect, while finite electrical resistivity has a destabilizing effect on the growth rate of the thermal instability. Routh-Hurwitz's criterion is used to calculate, the stability of the system. The results presented in this problem are helpful in understanding the molecular cloud formation and star formation in ISM.

Chapter 6 - Throughout this chapter, the authors have examined the effect of electron plasma frequency on gravitational instability in the magneto-radiation quantum plasma region. The research is carried out within the context of the normal mode analysis method, which has been improved due to the involvement of the electron plasma frequency. The dispersion relationship is simplified in the longitudinal and transverse propagation of the magnetic field. The expression instability is modified due to the existence of the magnetic field and the electron plasma frequency in the transverse mode of propagation. It is obvious from the curve that the electron plasma frequency has a destabilizing effect on the system. However, the presence of a quantum parameter reduces the detrimental effect of electron plasma frequency and stabilizes the system. The importance of the authors' research will allow us to better understand the stellar evolution of self-gravitational molecular clouds and the creation of stars.

In: An Introduction to Molecular Clouds ISBN: 978-1-53619-178-3
Editor: Sachin Kaothekar

Chapter 1

ROTATION OF DARK GLOBULES

A. L. Gyulbudaghian[*]
Victor Ambartsumian Byurakan Observatory

ABSTRACT

Dark globules have an important role in the evolution of the stars with low and/or middle masses. Many globules are connected with such stars, mainly with pre-main-sequence young stars (T Tauri type, Herbig A_e/B_e type et cet.). The investigation of rotation of the globules is important for clarifying the evolution of stars: as a rule, the specific angular momentum is much larger for the globules than for the binary and single stars. In this chapter we present the results of $^{12}CO(1\text{-}0)$ observations of several rotating dark globules. Angular velocities of investigated globules are in the range $(4 - 4.6)\cdot 10^{-14}$ s^{-1}, which is larger than the results obtained for the globules by other authors.

Keywords: dark globules, rotation, evolution of stars

---

[*] Corresponding Author's Email: agyulb@bao.sci.am.

1. INTRODUCTION

Since middle 70-es the molecular hydrogen became one of the main objects of the investigation of the physics of interstellar matter. The investigation of distribution of H_2 in the Galaxy reveals a discovery of new elements of the disc of our Galaxy: the molecular ring, which is a region of high concentration of H_2 in the ring with a radius of R = (4 -- 8) kpc from the center of our Galaxy. In this ring we have also a high concentration of stellar associations, HII regions, pulsars, SN remnants, the sources of diffuse gamma-radiation and synchrotron radiation. The molecular hydrogen is occurred mainly in the form of the molecular clouds which have different dimensions: the small clouds (globules), the middle size molecular clouds, the giant molecular clouds (GMC) and molecular complexes.

In this Chapter we will present the results of $^{12}CO(1\text{-}0)$ observations of three globules. These results are in favor of the existence of their rotation.

2. ROTATION OF AN ISOLATED DARK GLOBULE

This item is devoted to the investigation of an isolated small dark globule. In the central part of this globule an interesting CLN127-128 object is situated [1]: a system of three stars, almost on a straight line. Two stars are connected with bright nebulae, while the central star has no nebula. In [2] this object was interpreted as a central star with two lobes, but with I and R filters the three stars have the same images, so it is difficult to argue against their star origin. The central star has a spectrum of a B8e+K2 type star, the spectrum of southern star resembles the spectrum of an A0 type star. As the colors of northern and southern stars are similar, we can conclude, that the northern star also has an A0 spectrum.

An IRAS point source, IRAS 13542-6335, is connected with CLN127-128. The IRAS colors show, that the object IRAS 13542-6335 can be a T Tauri type star with a thick envelope [1].

The data of IR colors for northern and southern stars are rather different from zero, which is not common for a standard A0 type star, it means that northern and southern stars have thick envelopes.

The infrared colors of central star are much larger than the corresponding values for standard B8 or K2 type star, but are comparable with such values for Herbig A_e/B_e stars with thick shells or envelopes (like V380 Ori, PV Cep, HK Ori et cet.), therefore we can conclude that the central star is rather a Herbig A_e/B_e star than a double star [1].

The axis of the object (of three stars) lies at the position angle $34^0.0$ W of N. The three stars are collinear within $0^0.2$. The major axis of the globule is oriented by 18^0 from that of three-star system, which is rather close to each other.

We can suppose that here we have a group of three young stars, just emerged from the globule. Such groups of young stars, just emerged from the dark nebulae, were found also in other places. The existence of such groups is in favor of the following idea: many low-mass stars are born in groups. Such groups of low-mass stars are very unstable, because the gravitational attraction between the members is rather low (in the case of the systems consisting of high-mass stars the instability of the system is not so definite).

This globule was observed with the 15-m SEST (Swedish-ESO Submillimetre Telescope) telescope (Cerro La Silla, Chile) [1].

The telescope beam size at 115 GHz is 45” and the beam efficiency is 0.70. The positions towards CLN127-128 were observed with a spacing of 40” in the frequency-switched mode, in the frequency throw of 10 MHz. The telescope was equipped with SIS detector and a high-resolution acousto-optical spectrometer with 1000 channels and a velocity resolution of 0.112 km/s.

Table 1. Distribution of radial Velocity along the globule

200”					0”						200”	
			-22.0	-22.0								
	-22.5	-22.5	-22.5	-22.5	-22.0	-22.0						
-22.5	-22.5	-22.5	-22.5	-22.5	-22.5	-22.0	-22.0	-22.0	-22.0			
-21.4 -22.5	-21.4 -22.5	-21.4 -22.5	-22.0	-22.0	-22.0	-22.0	-22.0	-22.0	-22.0			
	-21.4 -22.5	-21.4 -22.5	-21.4 -22.5	-21.4 -22.5	**-22.0** *	**-22.0**	-22.0	-21.4				
	-21.4	-21.4 -22.5	-21.4 -22.5	-21.4 -22.5	-21.4 -22.5	-21.4	-21.4	-20.9				
	-21.4	-21.4	-21.4	-21.4	-21.4	-21.4	-21.4	-21.4	-21.4	-21.4		
		-21.4	-20.9	-20.9	-20.9	-20.9	-20.9	-20.9	-20.9	-21.4	=21.4	-22.0
			-20.4	-20.4	-20.4	-20.4	-20.4	-20.9	-20.9	-21.4	-21.4	-21.4
			-20.9	-20.9	-20.4	-20.4	-20.4	-20.4	-20.9	-20.9	-20.9	-20.9
					-20.4	-20.4	-20.4	-20.4	-20.4	-21.4	-21.4	-21.4
					-19.3	-19.3	-19.8	-19.8	-20.4	-21.4	-21.4	-21.4
					-19.3	-19.3	-19.8	-19.8	-19.8	-20.4	-20.4	
			-19.8	-19.8	-19.3	-19.3	-19.3	-19.3	-19.8	-20.4	-20.4	
			-19.3	-18.8	-18.8	-18.8	-18.8	-18.8	-19.3	-19.8		
				-18.8	-18.8	-18.8	-18.8	-18.2	-18.8	-19.3		
				-18.8	-18.8	-18.8	-18.2	-18.8	-19.3			

In Table 1, the distribution of $^{12}CO(1\text{-}0)$ velocity, obtained from observations, is given. The rows correspond to right ascension, increasing from top to bottom, and columns correspond to declination, increasing from right to left. The width of each column is 40", and the width of each row is also 40". The place of three stars is marked by an asterisk.

A velocity gradient is evident in the W – E direction, from -22.5 km/s to -18.2 km/s. As was confirmed by Ho and Therebey [3], a common rotating molecular cloud has the following characteristics: a clearly defined velocity gradient and also cloud's flat morphology, where the elongation is parallel to the velocity gradient, as to be expected in rotation. These characteristics of rotation take place also in the case of CLN 127-128: the globule is flattened, with the length to width ratio of 1.7.

There is an interesting phenomenon concerning this globule (see Table 1). From row 1 to row 7 the velocity of the globule material at the edges is more negative, than the velocity of the main body of the globule, while from row 7 to row 17 the velocity of the globule material at the edges is less negative than the velocity of main body of the globule. We can interpret this phenomena as follows. There is a rotation of globule as a whole around the axis of rotation, passing approximately through row 7. The material at the edges is less dense, and during the rotation the material is staying behind the main body of the globule. In the system connected with the globule we will have velocity, equal to 0 km/s near the axis of rotation of the globule, that is, the velocity of the globule in the LSR system will be -21.4 km/s. Table 2 shows the distribution of velocity in the system connected with the globule.

The distance to the globule is 1 kpc (Bruck and Godwin, [2]). We can calculate the momentum of inertia, angular velocity of rotation, specific angular momentum and energy of rotation of the globule.

Table 2. Distribution of radial velocity along the globule (In the LSR System)

			-0.54	-0.54								
	-1.1	-1.1	-1.1	-1.1	-0.54	-0.54						
-1.1	-1.1	-1.1	-1.1	-1.1	-1.1	-0.54	-0.54	-0.54	-0.54			
0.0 -1.1	0.0 -1.1	0.0 -1.1	-0.54	-0.54	0.54	0.54	0.54	0.54	0.54			
	0.0 -1.1	0.0 -1.1	0.0 -1.1	0.0 -1.1	-0.54 *	-0.54	-0.54	0.0				
	0.0	0.0 -1.1	0.0 -1.1	0.0 -1.1	0.0 -1.1	0.0	0.0	0.54				
	0.0	0.0	0.0	0.0	0.0	0.0	0.0	0.0	0.0	0.0		
		0.0	0.54	0.54	0.54	0.54	0.54	0.54	0.54	0.0	0.0	-0.54
			1.1	1.1	1.1	1.1	1.1	0.54	0.54	0.0	0.0	0.0
			0.54	0.54	1.1	1.1	1.1	1.1	0.54	0.54	0.54	0.54
					1.1	1.1	1.1	1.1	1.1	0.0	0.0	0.0
					2.1	2.1	1.6	1.6	1.1	0.0	0.0	0.0
					2.1	2.1	1.6	1.6	1.6	1.1	1.1	
			1.6	1.6	2.1	2.1	2.1	2.1	1.6	1.1	1.1	
			2.1	2.7	2.7	2.7	2.7	2.7	2.1	1.6	1.1	
				2.7	2.7	2.7	2.7	3.2	2.7	2.1		
				2.7	2.7	2.7	3.2	2.7	2.1			

The angular velocity of rotation of the globule $\Omega = \Delta V/\Delta R$, which means that it is equal to the gradient of the radial velocity. From the data, given above, the diameter of the globule will be about 640" (if the distance to the globule is 1 kpc, the diameter will be 3.2 pc), while the velocity difference for this size gives 4.3 km/s (from -1.1 km/s to 3.2 km/s). Hence we will have Ω = 4.3 km/s/3.2pc = $4.3\cdot 10^{-14}s^{-1}$. The period of rotation will be $T = 2\pi/\Omega = 4.7\cdot 10^6$ year. The momentum of inertia will be $I = (1/2)Mr^2$ (for the disc, as we supposed that the globule has a flat disc-like structure), where M is the mass of the globule and r is the radius of the disc. The mass of the globule was estimated in [2] as being 50 solar masses. We have $I = 2.1\cdot 10^{59} g^2cm^2\ s^{-2}$. The energy of rotation $W = (1/2)I\Omega^2 = (1/4)Mr^2\Omega^2$, hence $W = 10^{45}$erg. The angular momentum $J = I\Omega$, hence $J = 2.2\cdot 10^{22}$g $pc^2\ s^{-1}$. The specific angular momentum J/M = $2.2\cdot 10^{-13} pc^2\ s^{-1}$.

In [4] the rotation of globules, which have no embedded stars, is investigated. The specific angular momentum J/M of the starless globules ranges from $4\cdot 10^{-17}$ to $2\cdot 10^{-15}$ pc^2s^{-1} and is similar to the values for dense NH_3 cores in molecular clouds [5]. Simon et al. [6] made a near IR lunar occultation and direct imaging survey of T Tauri binary systems in the Ophiuchus and Taurus star-formation regions. The binaries observed ranged in separation from 3 AU to almost 1400 AU and on average contained 50 times less specific angular momentum, than the Goodman et al. [5] dense cores, or the Yun and Clemens [7] embedded proto stellar binaries ($10^{-15}pc^{-2}s^{-1}$), or the starless globules [4].

In our case the specific angular momentum of the globule ($2.2\cdot 10^{-13}pc^{-2}s^{-1}$) is much larger than in all cases described above. We can suspect that large amounts of specific angular momentum are being lost due to processes occurring after cloud core formation and fragmentation.

The $^{12}CO(1\text{-}0)$ observations suggest, that besides the rotation of the globule, a blue-shifted outflow from CLN 127-128 in the NE direction exists. The velocity of outflow is -22.5 km/s or, in the system, connected with the globule, is equal to -1.1 km/s.

3. Rotation of a Cloud Near RCW 38

The IR star cluster RCW 38 [8] is situated in a complex of molecular clouds with different velocities and sizes. $^{12}CO(1\text{-}0)$ observations show [9], that in the direction of RCW 38 exist at least three clouds. However, at the distance of RCW 38, 1.7 kpc, there are only two clouds, clouds 1 and 2 [9]. The object RCW 38 is situated in the depression in huge cloud 1 (the velocity of cloud 1 is about 0.54 km/s [9]). It is not excluded that this depression was formed by the object itself (by blowing away the material by stellar winds of young bright stars of star cluster RCW 38). The cloud 2 (the mean velocity is 5.4 km/s) is situated in the neighborhood of RCW 38 and has an elongation in SW-NE direction.

In Table 3, the distribution of $^{12}CO(1\text{-}0)$ velocity of cloud 2, obtained from observations [9], is given. The columns correspond to right ascension, increasing from right to left, while the rows correspond to declination, increasing from bottom to top. The width of each column is 40” and the width of each row is also 40”. The coordinates of center are: R.A.(2000) = $08^h59^m02^s.0$, DECL(2000) = $-47^0 29'43''$. From Table 3 is evident, that here exists a gradient of rotation. The gradient of velocity exists in SW-NE direction, from 4.05 km/s to 7.57 km/s. The existence of such a gradient means, that there is a rotation of cloud in SW-NE direction, with an axis of rotation having a SE-NW direction.

Table 3. Distribution of radial velocity along the Cloud 2

	80"	40"	0"	-40"	-80"
80"	7.57	7.03	5.95	5.68	---
40"	6.76	6.46	5.41	5.41	5.95
0"	5.68	6.22	5.41	5.14	4.87
-40"	4.87	5.68	5.40	4.59	4.33
-80"	4.33	5.14	4.59	4.33	4.05

The angular velocity of rotation of the cloud is $\Omega = \Delta V/\Delta R$, it means that it is equal to the gradient of the radial velocity. The gradient of rotation is maximal in SW-NE direction (see Table 3). Since the extension of cloud in that direction is 284" (from the Table 3), and because the distance to the object is 1700 pc, so its extension will be 2.41 pc. The velocity difference along the cloud in the SW-NE direction from the Table 3 is 3.52 km/s. Hence we have: $\Omega = \Delta V/\Delta R$ = 3.62 km/s/2.41 pc = $4.6\cdot 10^{-14}s^{-1}$. The period of rotation will be $T = 2\pi/\Omega = 4.4\cdot 10^{6}$year.

We can compare this value with the known so far values for other rotating objects. In [1] a rotating isolated globule is investigated. The angular velocity for that globule is $\Omega = 4.3\cdot 10^{-14}s^{-1}$. In paper [4] there are data on several rotating clouds (Bok globules) with angular velocities within the range $\Omega = (0.3 - 3)\cdot 10^{-14}s^{-1}$, it means that the cloud from this item has an angular velocity similar to the velocities of fast rotating globules.

There is a red-shifted outflow from the cloud, associated with RCW 38. The mean velocity of that outflow is equal to 11 km/s. The mean velocity of cloud is -5.4 km/s, so the outflow will be with velocity +5.6 km/s. This outflow is in SE direction from RCW 38, which is parallel to the axis of rotation of cloud and perpendicular to elongation of that cloud. If the cloud coincides with the dust disc around RCW 38, then the red outflow is perpendicular to the dust disc. Hence we can make a conclusion, that this phenomena is rather common for many YSO's (young stellar objects).

4. Rotation of a Part of Molecular Cloud, Connected with the Star CD -40 4427

This cloud is connected with an IR nebula, which has an IR star cluster, embedded in it [10]. From observations we can realize, that there are three emissions: emission from main cloud and emissions from red shifted and blue shifted outflows, initiated by a double star [10]. The main cloud velocity is 7.73 km/s, the velocity of red shifted outflow is 9.001 km/s (or +1.271 km/s in respect with the velocity of main cloud), the velocity of blue shifted outflow is 5.910 km/s (06 - 1.820 km/s in respect with the velocity of main cloud).

Below is presented Table 4, in which the velocities (in km/s) of main cloud are given.

Table 4. Distribution of velocity of main Cloud

	80"	40"	0"	-40"	-80"
80"	7.73	7.548	7.548	7.73	7.73
40"	7.73	7.184	7.548	7.912	7.912
0"	7.184	7.184	7.366	7.548	7.73
-40"	7.002	6.82	7.366	7.366	7.73
-80"	6.638	6.82	7.184	7.366	7.73

We can see from Table 4, that the eastern part of the region observed in $^{12}CO(1\text{-}0)$ is rotating, because there is a gradient of velocity in N – S direction. On the length of 160" (which at the distance of 1.1 kpc corresponds to $2.31\cdot10^{13}$km) the gradient of velocity is equal to 1.092 km/s. Hence the angular velocity of rotation $\Omega = \Delta V/\Delta R = 1.092\ \text{km/s}/2.31\cdot10^{13}\text{km} = 4.3\cdot10^{-14}\text{s}^{-1}$. We can obtain also the period of rotation: $T = 2\pi/\Omega = 1.44\cdot10^{14}\text{s} = 4.64\cdot10^{6}$year.

Rotation of eastern part of the region is around the axis of rotation, which has an E – W orientation. The eastern part of the region coincides with an IR nebula, and the northern part of that nebula has also an E – W orientation. Hence we can make the following conclusion: the northern part of the nebula can also rotate around the same axis of rotation. If so, it is not excluded, that the northern part of IR nebula has a prolongation along E – W direction.

CONCLUSION

In this chapter the results of the $^{12}CO(1\text{-}0)$ observations of three molecular clouds are presented.

The first object is an isolated globule, connected with a chain of three young stars. The $^{12}CO(1\text{-}0)$ observations of the globule revealed the existence of a velocity gradient along the globule, which implies the presence of rotation of the globule as a whole with an angular velocity equal to $4.3 \cdot 10^{-14} s^{-1}$. We could establish the place of the axis of rotation by analyzing the complex structure of rotation: at the edges of the globule the material is staying behind the main body of the globule. In the part approaching us the velocity of that material is larger, than the velocity of main body of the globule, while in the part moving away from us, the velocity is smaller. These observations also suggest the existence of a blue-shifted outflow in the NE direction from the CLN 127-128 (a chain of three stars, connected with the globule), with outflow velocity -1.1 km/s (in the system, connected with the globule).

The second cloud is connected with IR star cluster RCW 38. $^{12}CO(1\text{-}0)$ observations showed the existence of several molecular clouds, two of which are associated with RCW 38. Cloud 1 is a larger one with a depression, in which the cluster might be embedded (it is not excluded that the depression was formed by the young bright stars of RCW 38). Cloud 2 is a bar-shape or disc-shape cloud, which is rotating around the axis of rotation with an angular velocity $4.6 \cdot 10^{-14} s^{-1}$, so it is a rather fast rotator. A red-shifted molecular outflow in SE direction from RCW 38 (with velocity +5.6 km/s) was discovered. That outflow is perpendicular to the elongation of cloud 2.

The third cloud is around an unknown IR nebula. $^{12}CO(1\text{-}0)$ observations revealed existence of red-shifted and blue-shifted outflows around a double star with nebular tails. The rotation of a part of dark cloud was also discovered, with an angular velocity equal to $4.3 \cdot 10^{-14}$ s^{-1}. This rotation is supposed to be a part of rotation of IR nebula.

References

[1] Gyulbudaghian, A. A., May, J. (2004). On rotation of an isolated globule. *Astrophysics*, 47, 353-360.

[2] Bruck, M. T., Godwin, P. J. (1984). Spectroscopic and photometric observations of an unusual bipolar nebula in Bok globule. *MNRAS*, 206, L37-L46.

[3] Ho, P. T .P., Terebey, S., Turner, J. L. (1994). The rotating molecular core in in G10.6-0.4. *ApJ*, 423, 320-332.

[4] Kane, B. D., Clemens, D. P. (1997). Rotation of starless Bok globules. *AJ*, 113, 1799-1814.

[5] Goodman, A. A., Benson, P. J., Fuller, G. A., Myers, P. C. (1993). Dense cores in dark clouds. VIII. Velocity gradients. *ApJ*, 406, 528-535.

[6] Simon, M., Ghez, A. M., Leinert, Ch. et al. (1995). A lunar occultation and direct imaging survey of multiplicity in Ophyuchus and Taurus star forming regions. *ApJ*, 443, 625-637.

[7] Yun, J. L., Clemens, D. P. (1994). Near-Infrared imaging of young stellar objects in Bok globules. *AJ,* 108, 612-623.

[8] Bica, E., Dutra, C. M., Barbuy, B. (2003). A catalog of infrared star clusters and stellar groups. *A&A*, 397, 177-180.

[9] Gyulbudaghian, A. L., May, J. (2008). Investigation of an conspicuous infrared star cluster and star-forming region "RCW 38 IR cluster". *Astrophysics*, 51, 18-28.

[10] Gyulbudaghian, A. L., Mendez, R. (2014). New radial systems of dark globules and HH objects. *Astrophysics*, 57, 520-529.

In: An Introduction to Molecular Clouds ISBN: 978-1-53619-178-3
Editor: Sachin Kaothekar

Chapter 2

A NEW SIMPLISTIC EQUATION OF A MACROSCOPIC EQUILIBRIUM STATE FOR DEGENERATE NEUTRON STARS

Rohaan Deb[1] and Pralay Kumar Karmakar[2,*]
[1]Department of Physics, Rheinische Friedrich-Wilhelms-Universität Bonn, Bonn, Germany
[2]Department of Physics, Tezpur University, Napaam, Tezpur, Assam, India

ABSTRACT

A new simplistic model equation of macroscopic equilibrium state for understanding the realistic gyro-equilibrium properties of degenerate neutron stars is semi-analytically developed. The adopted model setup prudently includes all the possible realistic active tonality agencies unavoidable in the course of realistic stellar fluid dynamics concurrently. As such, it conjointly includes thermal pressure, quantum degeneracy pressure, turbulence pressure, self-gravity, magnetic field, viscosity, and the Coriolis rotational effects on the relevant scales. A systematic combination of the fundamental and derived multi-parametric analyses renders the obtained resultant equation of macroscopic state in a standard

[*]Corresponding Authors Email: pkk.766@live.com.

normalized form of an ordinary differential (first-order) equation on the material density alone. It is numerically solved, emphatically analyzed and graphically illustrated. A direct non-trivial correlation of the density on the explicit dependencies on temperature, viscosity and electric current is portrayably established. The expounded semi-analytic findings match fairly with the existing non-homologous predictions extensively available in the literature.

Keywords: stellar structure, interiors, evolution, nucleosynthesis, ages, neutron stars, kinetic theory

1. INTRODUCTION

This is a well-established fact that neutron stars are compact astrophysical objects having extremely strong magnetic fields (B~10^{13} G). They are gravitationally condensed structures having a strong inward self-gravity with the outward support provided by degeneracy pressure [1-3]. The theoretic framework of polytropic equation of state, although based only on pure gravito-thermal interplay, has widely been employed in describing the degenerate neutron-rich matter constituting such stellar structures. It ignores a significant number of key factors responsible for the dynmical tonality of suggestors. The compositional matter forming neutron stars exhibits a wide plethora of extreme behaviours [4-9]. It includes superfluidity, superconductivity (T_c~10^{10} K), and so forth. Besides, various invariant collective dynamics evolving in the form of waves, instabilities and fluctuations in such fluid media are yet to be well explored [5-10]. It can, thus, be seen that there has been a long-sought plea for a simplified equation of state depicting at least the hydrodynamic perspective of neutron stars and like objects in a closed form [7-11].

In this Letter, we procedurally develop a simplistic semi-classical equation of macroscopic equilibrium state with a view to fully characterize the degenerate matter in the presence of all the possible realistic active agencies unavoidable in the course of realistic fluid dynamics associated with neutron stars. In a wider horizon, it considers the pressure due to: (a) Coriolis force, (b) magnetic force, (c) viscous force, (d) degeneracy force of quantum-mechanical origin, (e) gravitatonal force, (f) fluid turbulence force and (g) thermal force simultaneously. The net force-balancing condition leads to a unique type of first-order ordinary differential equation on the degenerate material density. It is herewith expounded semi-analytically to validate the applicability of the results in the realistic compact astrophysical context of active evolution.

2. Physical Model and Calculation

We consider a neutron star in hydrostatic equilibrium in a spherically symmetric (1-D) configuration space. The supports are primarily provided by the quantum degeneracy and gravitational pressures. In this formalism, we consider a number of stellar realistic parametric factors acting simultaneously, such as turbulence, viscosity, degeneracy, Coriolis rotational effect, thermal pressure, and magnetic field. To be more specific, it may be mentioned here that pressure due to turbulence, Coriolis effect, thermal factor, degeneracy, and magnetic field operate outward relative to the centre of the neutron star; whereas, the pressure due to viscosity and self-gravity act inward. The combined action of all the considered agencies establishes a new form of hydrostatic equilibrium in the neutron star. In order to model the macroscopic state, an elementary concentric spherical shell of radius r and radial thickness dr is considered inside the star. Thus, the equilibrium setup of the model configuration in a force-balanced condition [3-8] in an inertial frame of reference can elementarily be expressed as

$$P_{turbulence} + P_{Coriolis} + P_{\deg eneracy} + P_{thermal} + P_{magnetic} = P_{viscous} + P_{gravity} \quad (1)$$

where, $P_{turbulence}$, $P_{Coriolis}$, $P_{degeneracy}$, $P_{thermal}$, $P_{magnetic}$ $P_{gravity}$ and $P_{viscous}$ are the active forces sourced by the fluid turbulence, Coriolis rotational effects, quantum degeneracy effect, thermal pressure, magnetic effects, self-gravity and fluid viscosity; respectively.

Let us now obtain the explicit expressions for the various component pressures constitutively appearing in eq. (1) as given in the following. We use the logatropic equation of state [4] relating the isothermal gas pressure (linear) and non-isothermal turbulence pressure (non-linear) as a first step. It may be note worthy here that the logatropic equation of state is valid only in the case of isolated star formation and static situations occurring in gaseous molecular clouds as in our present case. The logatropic equation of state of our concern involving turbulence pressure, which is an empirically obtained mathematical relationship after astrophysical spectral line-width variations [5], can explicitly be written in the customary notations as

$$P_t = P_0\left[1 + A\ln\left(\frac{\rho}{\rho_0}\right)\right], \tag{2}$$

where, A is a constant (A~0.2 ± 0.02), $\rho_0 = n_0 m$ is a reference (equilibrium) material density and P_0 is a reference (equilibrium) pressure for the stellar fluid composed of constituent identical particles (neutron) of mass m each having population density n_0 at the equilibrium temperature T K. Now, the expression for the pressure due to the force of gravity in accordance with the Newton law [1,2] is written as,

$$P_{gravity} = \frac{\rho g}{4\pi r^2}, \tag{3}$$

where, ρ is the fluid material density, g is the acceleration due to gravity and r is the radius of the spherical shell as considered above. Likewise, the pressure due to Coriolis effect(which in principle is an inertial force that acts on objects that are in motion relative to a rotating frame of reference) experienced by the structure moving with a mean linear velocity v and angular velocity Ω can be expressed as

$$P_{Coriolis} = \frac{-2\rho\left|\left(\vec{\Omega}\times\vec{v}\right)\right|}{4\pi r^2} = \frac{-2\rho\Omega v\sin\theta}{4\pi r^2}, \tag{4}$$

where, θ is the angle between the vectors $\vec{\Omega}$ and $\vec{v}$, here θ being considered as constant for the simplicity of the calculation. Now, applying the formula $v = \sqrt{2GM/r}$ for an analytic simplification for the description of the bulk macroscopic state of the stellar configuration, eq. (4) can be rewritten in an explicit mathematical form as

$$P_{Coriolis} = \frac{2\sqrt{2}\Omega G^{1/2} M^{1/2}\sin\theta}{4\pi}\left(\frac{\rho}{r^2}\right), \tag{5}$$

where, all the notations used throughout are generic and they all carry usual significances. Again, the expression for the degeneracy pressure is given as,

$$P_{\deg eneracy} = \frac{\left(3\pi^2\right)^{2/3} h^2}{5m}(\rho)^{5/3}, \tag{6}$$

where, h is the Planck constant (=6.634×10^{-34} J s). Now, the thermal pressure inside the neutron star shell is cast as,

$$P_{thermal} = \frac{1}{3}\rho c^2, \tag{7}$$

where, c is the speed of light. Applying the expression $c^2 = (3\pi/8)v^2$, where, $v = \sqrt{8RT/\pi m}$ [6], eq. (7) can be rewritten as

$$P_{thermal} = \frac{RT}{m}\rho. \tag{8}$$

Likewise, the expression for the pressure due to the magnetic field B of the neutron star of radius R is expressed as

$$P_{magnetic} = \frac{B}{2\mu_0}. \tag{9}$$

Now, applying the justified approximation of $B = \mu_0 IR^2 / 2(R^2 + r^2)^{3/2}$ in eqn. (9), one gets

$$P_{magnetic} = \frac{\mu_0 I^2 R^4}{8(r^2 + R^2)^3}. \tag{10}$$

Lastly, the pressure due to viscous drag force arising due to internal atomic or molecular friction at a collective level is as

$$P_{viscous} = -\eta \frac{dv}{dr}, \tag{11}$$

where, η is the fluid viscosity. Now, using the relation $v = (2GM/r)^{1/2}$; where, G is the universal gravitational constant and M is the source mass, i.e., mass of the inner sphere, eq. (11) can be rewritten as

$$P_{viscous} = \left(\eta^2 GM/2\right)^{1/2} (r)^{-3/2}. \tag{12}$$

Now, applying all the derived pressure expressions from above equations (eqs. (2)-(3), eqs. (5)-(6), eqs. (8), eqs. (10) and eqs. (12)) and simplifying eq. (1), one finally obtains

$$\begin{aligned}&\left(a_1 + a_2\rho^{2/3} + a_3\rho^{-1} - a_4 r^{-5/2} - a_5 r^{-2}\right) d\rho/dr + \\ &\left(a_6 r^{-7/2} + a_7 r^{-3}\right)\rho + a_8 r^{-5/2} - a_9\left(r/\left(R^2 + r^2\right)^4\right) = 0,\end{aligned} \tag{13}$$

where,

$$\begin{aligned}&a_1 = RT/m,\ a_2 = \left(3\pi^2\right)^{2/3} h^2/3m,\ a_3 = P_0 A\rho_0, \\ &a_4 = 2\sqrt{2}\Omega G^{1/2} M^{1/2} \sin\theta/4\pi,\ a_5 = g/4\pi, \\ &a_6 = 5\sqrt{2}\Omega G^{1/2} M^{1/2} \sin\theta/4\pi,\ a_7 = g/2\pi, \\ &a_8 = 3/2\left(\sqrt{\eta^2 GM/2}\right),\ a_9 = 3\mu_0 I^2 R^4/4.\end{aligned}$$

The focal goal of our analysis lies in characterizing the considered neutron star configuration in a standard scale-free invariant form. Accordingly, we transform all the relevant physical parameters (dimensional) in a normalized (dimensionless) form by using a standard astrophysical normalization scheme [5,7,9,10]. The new set of parameters is described as $\xi := r/\lambda_J$ for the normalized distance and $\rho^* := \rho/\rho_0$ for the normalized pressure. Here, $\lambda_J = \left(kT/4\pi Gm\rho\right)^{1/2}$ is the well-known Jeans scale length. Therefore, eq. (13) in the normalized form with these reservations can be written as

$$\left[b_1T^{-1/2}+b_2T^{-1/2}\left(\rho^*\right)^{2/3}+b_3T^{-1/2}\left(\rho^*\right)^{-1}-b_4T^{-7/4}\xi^{-5/2}-b_5T^{-3/2}\xi^{-2}\right]$$
$$\frac{d\rho^*}{d\xi}+\left[b_6T^{-7/4}\xi^{-7/2}+b_7T^{-3/2}\xi^{-3}\right]\rho^*+b_8T^{-5/4}\xi^{-5/2}$$
$$-b_9T^{1/2}\xi=0,$$

$$b_1=a_1(\rho_0)^{3/2}\left(4\pi mG/k\right)^{1/2}, b_2=a_2\left(\rho_0\right)^{5/2}\left(4\pi mG/k\right)^{1/2}, \quad (14)$$
$$b_3=a_3\left(\rho_0\right)^{1/2}\left(4\pi mG/k\right)^{1/2}, b_4=a_4\left(\rho_0\right)^{11/4}\left(4\pi mG/k\right)^{1/2},$$
$$b_5=a_5\left(\rho_0\right)^{5/2}\left(4\pi mG/k\right)^{1/2}, b_6=a_6\left(\rho_0\right)^{11/4}\left(4\pi mG/k\right)^{1/2},$$
$$b_7=a_7\left(\rho_0\right)^{5/2}\left(4\pi mG/k\right)^{1/2}, b_8=a_8\left(\rho_0\right)^{5/4}\left(4\pi mG/k\right)^{1/2},$$
$$b_9=a_9\left(\left(\rho_0\right)^{-1/2}/R^3\right)\left(k/4\pi Gm\right)^{1/2}$$

After a little bit of algebraic simplification, eq. (13) describing the equilibrium dynamics of the spatially evolving stellar structure on the Jeans scale of space reduces to

$$\frac{d\rho^*}{d\xi}+A_1(\rho^*,\xi,T)\rho^*+A_2(\rho^*,\xi,T)=0, \quad (15)$$

where, $A_1(\rho^*,\xi,T)=\left(b_6T^{-7/4}\xi^{-7/2}+b_7T^{-3/2}\xi^{-3}\right)/$
$\left(b_1T^{-1/2}+b_2T^{-1/2}\left(\rho^*\right)^{2/3}+b_3T^{-1/2}\left(\rho^*\right)^{-1}-b_4T^{-7/4}\xi^{-5/2}-b_5T^{-3/2}\xi^{-2}\right)$
and $A_2(\rho^*,\xi,T)=\left(b_8T^{-5/4}\xi^{-5/2}-b_9T^{1/2}\xi\right)/$
$\left(b_1T^{-1/2}+b_2T^{-1/2}\left(\rho^*\right)^{2/3}+b_3T^{-1/2}\left(\rho^*\right)^{-1}-b_4T^{-7/4}\xi^{-5/2}-b_5T^{-3/2}\xi^{-2}\right)$.

Thus, we can say that eq. (15) represents a simplistic form of the generalized equation of state for the family of neutron stars and live stellar structures. Clearly, it is in a normalized differential (first-order) form and conveniently analyzable.

3. RESULTS AND DISCUSSIONS

As a consequence of the mathematical analyses, a new equation of state (eq. (15)) for theoretically characterizing the equilibrium structure of complex neutron star configurations in a normalized (scale-free) form is analytically developed. An exploration for the explicitanalytical solutions, be it approximate, is found to be tedious. A numerical analysis for investigating the exact bounded solutions with the help of the fourth–order Runge-Kutta method [13] is constructed. In order to characterize the average dynamical features, we judiciously choose [1-3] the quantitative values of the different multi-parametric coefficients in eq. (11) as, $b_2 = 10^5$, $b_3 = 10^{-15}$, $b_4 = 10^{13}$, $b_5 = 10^{-14}$, $b_6 = 10^{-13}$, $b_7 = 10^5$, $b_8 = 10^{10}$, $b_9 = 10^{12}$. The initial seed value of the pressureevolution is taken as $\rho_0^* = 10^{-3}$. The numerical results thus obtained by the above integration technique (RK-IV method) are graphically displayed in Figures1-3.

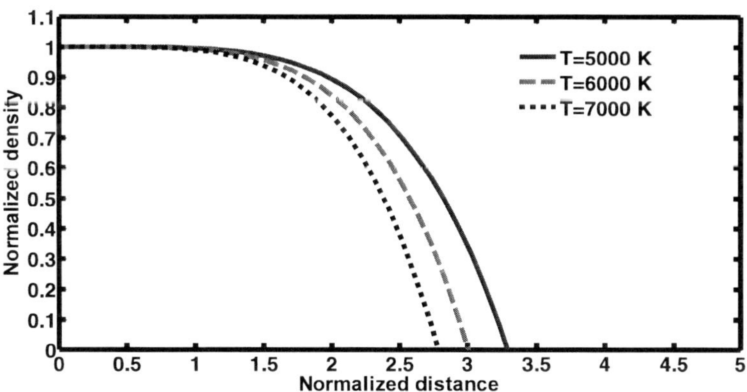

Figure 1.Spatial profile of the normalised density (ρ^*)for different values of temperature (T), but with a fixed viscosity ($\eta = 10^{30}$ N s m^{-2}) and fixed current (I = 10^{30}A) . The various lines correspond to T = 5000 K (blue solid line), T = 6000 K (red dashed line), and T = 7000 K (black dotted line); respectively. The fine details are in the text.

In Figure 1, we portray the spatial variation of the normalized neutron star density of material distribution as a hydrostatically bounded configuration for three different values of the stellar temperature, but with a fixed viscosity ($\eta = 10^{30}$ N s m^{-2}) and fixed current (I = 10^{30}A). The various lines correspond to T = 5000 K (blue solid line), T = 6000 K (red dashed line), and T = 7000 K (black dotted line); respectively. It is seen that the effective normalized densitygoes on decreasing as we move further from the centre of the neutronstar mass distribution, i.e., it asymptotically decreases to some non-zero finite value from the centre outwards. It is further found that, more the stellar temperature, less is the localised density; and vice-versa. Almost similar features, as Figure1, are found to exist even in Figure2, but for different current values with a fixed temperature,T = 6000 K and fixed value of viscosity, $\eta = 10^{30}$ N s m^{-2}. The various lines now correspond to I = 10^{30} A (blue solid line), I = 10^{31} A (red dashed line), and I = 10^{32} A (black dotted line); respectively.

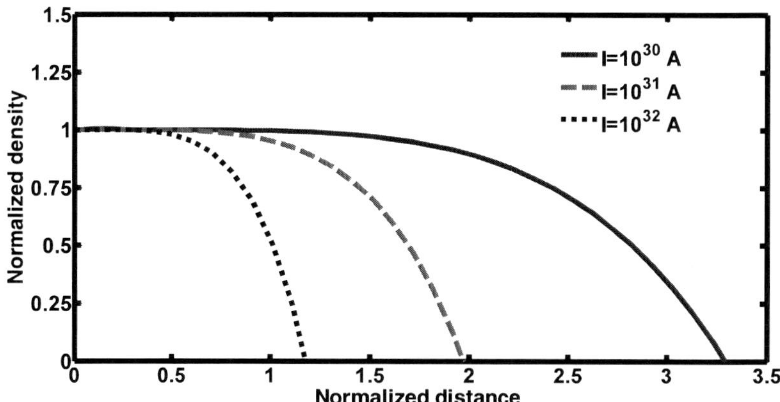

Figure 2: Same as Figure 1, but for different current values with fixed temperature T = 6000 K and fixed viscosity $\eta = 10^{30}$ N s m^{-2}. The various lines now correspond to I = 10^{30} A (blue solid line), I = 10^{31} A (red dashed line), and I = 10^{32} A (black dotted line); respectively.

It is speculated that the effective localized normalised density increases with decrease in current, and vice-versa. In both the cases (Figures 1-2), as confirmed elsewhere [6, 11,12] too, the maximization of the effective density is attributable to the nonlocal self-gravitational condensation for bounded neutron stars to exist dynamically evolving.

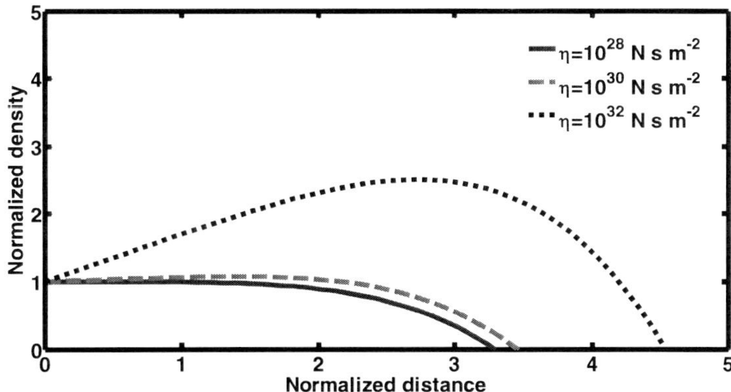

Figure 3. Same as Figure 1, but for different viscosity values with fixed temperature T = 6000 K and fixed current, I = 10^{30} A. The various lines now correspond to $\eta = 10^{28}$ N s m^{-2} (blue solid line), $\eta = 10^{30}$ N s m^{-2}(red dashed line), and $\eta = 10^{32}$ N s m^{-2}(black dotted line); respectively.

Lastly, by varying the viscosity, keeping the electric current and temperature constant, we obtain a set ofanalogous profiles as in Figure3.Unlike the previous cases with thhe increase inviscosity (Figure 2),now there is an increase in the normalized effective density including the sudden increment of the value ofeffective density with viscosity above 10^{32}N s m^{-2} (Figure3).The profile patterns obtained here on the basis of the derived equation of state are interestingly found to be in good conformity with the previously reported results [6]. Thus, we can say that the basic properties of neutron stellar structures can be studied with the help of the afresh derived equation of state (eq. (15)) in normalized differential form.

CONCLUSION

In conclusive summary, a new model equation of macroscopic state to investigate the gyro-equilibrium dynamical properties of neutron stars in the presence of various unavoidable realistic parametric factors with respect to a statical inertial frame of referenceis theoritically devised and semi-analytically explored.The adopted semi-classical model considers the conjoint balanced action of all the multi-parametric fluid-functional agencies, such as pressures due to viscous force (inward), gravitational force (inward), fluid turbulence force (outward), degeneracy effect (outward), magnetic force (outward), and Coriolis force (outward) along with thermal pressure (outward) relative to the centre of the neutron star mass distribution concurrently. The resulting new dimensional equation of state is transformed into a standard normalised (dimensionless) form for a scale-free description of the stellar equilibrium characteristic features. Analytical complications involved in the mathematical construct of the methodoligically derivedequation opens a new challenge for finding explicit bounded solutions enabling the study of proper behaviour of a neutron star. We, therefore, construct a detailed numerical illustrative scheme to obtain the exact solutions thereof, via the fourth-order Runge-Kutta method [13].

The analysis gives a direct corelation of the net effective density in the parametric space defined by the stellar temperature, viscosityand current; and so forth. The practical reliability of the investigated results as special astro-corollaries is examined and validated in the panoptic light ofthe previous conformal predictions for bounded polytropic stellar configurations reported elsewhere [6-12].

ACKNOWLEDGMENT

It is admitted herewith that the active cooperation received from Tezpur University is to be thankfully accredited. We are equally appreciative to the Editorial Board Members. At the last, the financial support received (by the communicating author) through the SERB Project (Grant- EMR / 2017 / 003222) is appropriately recognized.

REFERENCES

[1] Chandeasekhar, S, (1939). *An Introduction to the Study of Stellar Structures* (Dover).

[2] Lattimer, M. & Prakash M., (2016). "The equation of state of hot, dense matter and neutron stars", *Phys. Reports* 621:126-164.

[3] Kippenhahn, R., Weigert, A. & Weiss A., (2012). *Stellar Structures and Evolution* (Springer, London).

[4] Schmidt, W., (2014). *Numerical Modelling of Astrophysical Turbulence* (Springer, London).

[5] Borah, B. & Karmakar P. K., (2015). "A theoretical model for electromagnetic characterization of a spherical dust molecular cloud equilibrium structure", *New Astron.* 40:49.

[6] Hunter, C., (2001). "Series solutions for polytropes and the isothermalsphere" *Mon. Not. Royal Astron. Soc.*, 328:839.

[7] Chandrasekhar, S., (1951). "The gravitational instability of an infinitehomogeneous turbulent medium", *Proc. Royal Soc. London. Ser. A* 210:26.

[8] Chandrasekhar, S. & Fermi, E., (1953). "Problems of gravitationalstability in the presence of a magnetic field", *Astrophys. J.*, 118:116.

[9] Vazquez-Semadeni, E., Canto, J. & Lizano, S., (1998). "Doesturbulent pressure behave as a logatrope?", *Astrophys. J.*, 492:596.

[10] Maeder, A., (2009). *Physics, Formation and Evolution of Rotating Stars* (Springer, Heidelberg).

[11] Bonnano, A., Baldo, M., Burgio, G. F.& Urpin, V., (2014). "The neutron star in Cassiopeia A: equation of state, superfluidity, and Joule heating", *Astron. Astrophys.* 561:L5.

[12] Lindfield, G. R. & Penny, J. E. T., (2012). *Numerical Methods Using MATLAB* (Elsevier, UK).

In: An Introduction to Molecular Clouds ISBN: 978-1-53619-178-3
Editor: Sachin Kaothekar

Chapter 3

A NEW *P(R)*-RELATIONSHIP FOR STELLAR STRUCTURES

Rohaan Deb[1] and Pralay Kumar Karmakar[2,*]
[1]Department of Physics, Rheinische Friedrich-Wilhelms-Universität Bonn, Bonn, Germany
[2]Department of Physics, Tezpur University, Napaam, Tezpur, Assam, India

ABSTRACT

A new barotropic model equation of the macroscopic equilibrium state for understanding the gyro-equilibrium properties of spherically symmetric stellar structures is semi-analytically developed. The adopted model setup judiciously includes all the possible realistic stellar parametric factors concurrently, such as non-local self-gravity, fluid turbulence, viscosity, Coriolis rotational effects and relevant radiative hydrodynamical effects. The obtained resultant radial differential equation of state in a standard normalized form is numerically solved as an initial value problem and portraylly illustrated. Despite the accounted complications, a direct correlation of the net pressure composed of both

[*]Corresponding Author's Email: pkk.766@live.com.

the material and radiative hydrodynamic contributions is analyzed and established. The expounded semi-analytic findings fairly match with the existing predictions available in the literature as special astro-corollaries. The proposed analysis can be useful to study various thermophysical features of the stars and their ambient atmospheres.

Keywords: equations of state of non-metals, stellar structure, interiors, evolution, nucleosynthesis, age, kinetic theory

1. Introduction

It is well known that one of the most widely employed equations of state describing the stellar structures is the polytropic equation of state. A polytrope is a self-gravitating sphere comprising of the stellar material fluid based on a crude approximation of gravito-thermal force balancing with a number of important fluid dynamical properties remaining completely ignored therein [1-2]. It is however appreciable that the polytropic treatment has been fairly successful in depicting the behavior of the stellar material concentration as being highly weighted towards the center of the stellar structures thereby ensuring the well-established universal principle of mass conservation [2]. Moreover, an appropriate application of such equations of state, for pure field-free descriptions of dynamics in unsophisticated configurations, has been found to be extensively and commodiously useful in scientific description of a wide class of stellar, astrophysical and cosmological structures in different workspaces [1-8].

In the recent past, a full electromagnetic characterization of interstellar spherical dust molecular clouds, giving birth to stars and like astrophysical bounded structures in the absence of fluid turbulence [3] has been systematically carried out by applying the polytropic model hypothesis [4]. It has legitimately been successful in empirically determining the lowest-order cloud surface boundary formed due to gravito-electrostatic interplay. All the related electromagnetic cloud properties as a dynamically coupled system in a closed form have also been reported. It may be noted here that the polytropic equations of state holds good only for a system in hydrostatic equilibrium in the absence of a number of significant factors associated with the stars and the intervening matter [5]. The polytropic formalism would indeed deviate towards hydrodynamic equilibrium, especially, in the presence of fluid turbulence, viscosity, Coriolis force, etc. It is also evident from a continuous array of research and simulations that, the process of star formation is very inefficient in nature which is clear from the fact that, in spite of the presence of billions of solar masses of fresh gas from stars, it produces only about one solar mass of new stars per year. Thus, we can say that there must be some process in play which might be hindering the core of dense stellar gas from collapsing. This slow and inefficient formation of stars in the universe can be explained using two important phenomenons, namely, magnetic field and turbulence. However, the magnetic force is not enough to balance out the gravitational collapse because if it did then there would not have been any process of formation of star to begin with. Thereby, we are finally left with turbulence in order to explain this anomaly. It has been furthermore seen that the kinetic energy carried by fluid turbulence is comparable to the energy carried by the gravitating cloud fluids [3].

In this brief Letter, we report a strategic development of a new model radial barotropic equation of bulk macroscopic state for characterizing stellar structures in the presence of all the possible realistic active agencies unavoidable in realistic stellar fluid dynamics. In a wider horizon, that is, it considers the Coriolis rotation force, viscous force arising from microscopic particle resistive collisional effects, force due to gravity and the force sourced by fluid turbulence simultaneously for the first time. Moreover, the first principle of radiation hydrodynamics, which takes care of the modification of the fluid dynamical equilibrium with the help of electromagnetic ray absorption-emission processes by moving fluid [3], is also used. The analysis is expounded both analytically and numerically to validate its realistic applicability in astronomical environments with a particular emphasis to structuralize dust molecular clouds bringing about bounded structures in the light of existing fluid accretive-decretive scenarios.

2. Physical Model and Calculation

We consider, as a primitive imagination, a typical stellar system in hydrostatic inhomogeneous equilibrium ($d/dr \neq 0, d/dt \sim 0$) in one-dimensional (1-D) configuration space in the absence of any kind of external influences from electromagnetic fields, gravitational fields (tidal effects) sourced from any kind of distant astrophysical objects, etc. A number of realistic possible parametric factors acting simultaneously are considered afresh. It carefully includes fluid turbulence (nonlinear barotropic effects), dynamical viscosity, self-gravitating and Coriolis rotational effects concurrently. It may be mentioned here that the force due to self-gravity and that due to viscosity act inward; whereas, those due to the turbulence and Coriolis effects operate outward with respect to the centre of the entire stellar matter mass distribution. The co-occurring conjoint action of all these deliberated agencies establishes a new organized form of hydrostatic equilibrium yet to be seen.

In order to model the macroscopic unperturbed state of the stellar structure, an elementary concentric spherical shell of radius r and radial thickness dr is imagined. The equilibrium setup condition for the model configuration in an inertial frame of reference can elementarily be configured as

$$F_{turbulence} + F_{Coriolis} + F_{rad} = F_{gravity} + F_{viscous}, \tag{1}$$

where, $F_{turbulence}$, $F_{Coriolis}$, $F_{gravity}$ and $F_{viscous}$ are the active forces sourced by the fluid turbulence (outward), Coriolis rotational effects (outward), self-gravity (inward) and fluid viscosity (inward); respectively. Gravity (inward) and fluid viscosity (inward), respectively. The choice of the operational sense of the different forces is judiciously based on the star-centric and anti-star-centric force action law [11-14].

Let us now obtain the explicit expressions for the various component forces appearing in eq. (1) as in the following. We use the non-linear logatropicbarotropic equation of state [11-14], relating the isothermal gas pressure (linear on density) and non-isothermal turbulence pressure (non-linear on density) as a first step. It may be admitted here that the logatropic equation of state is valid only in the case of isolated star formation and static situations occurring in gaseous molecular clouds as in our present case. The logatropic equation of state of our concern involving fluid turbulence pressure, which is an empirically obtained mathematical relationship after astrophysical spectral line-width variations [3], can explicitly be written in the customary notations as

$$P_t = P_0\left[1 + A\ln\left(\frac{\rho}{\rho_0}\right)\right], \tag{2}$$

where, A is a constant (A~0.2 ⊥ 0.02), $\rho_0 = n_0 m$ is a reference (equilibrium) material density, and P_0 is a reference (equilibrium) pressure for the stellar fluid material composed of constituent identical particles of mass m each having population density n_0 at an equilibrium temperature T K given with the help of the well-known Virial theorem [3] as

$$P_0 = n_0 kT = \frac{Gm^2}{5r}. \tag{3}$$

Now, eq. (2) on spatial differentiation once, yields

$$\frac{dP_t}{dr} = \frac{d}{dr}\left[P_0\left[1 + A\ln\left(\frac{\rho}{\rho_0}\right)\right]\right] \tag{4}$$

The kinetic force exerted due to the fluid turbulence can be obtained from eqs. (3)-(4) in a compact simplified form as

$$F_{turbulence} = \frac{dP_t}{dr} = -\frac{P}{r} + \frac{8\pi}{m}\cdot\rho\,\mathrm{Pr}^2 + \frac{GAm^2}{5}\cdot\frac{1}{\rho r}\frac{d\rho}{dr}, \tag{5}$$

where, $G = 6.67\times10^{-11}$ $\mathrm{m}^3\mathrm{kg}^{-1}\mathrm{s}^{-2}$ is the universal gravitational constant via which self-gravitational interaction of matter is realized. Here, since pressure due to turbulence has a significantly large contribution, so we consider $P \sim P_t$ in eq. (5). Now, the expression for the force due to gravity in accordance with the Newton law [3] can be cast as

$$F_{gravity} = \rho g, \tag{6}$$

where, ρ is the fluid material density and g is the acceleration due to gravity. Likewise, the magnitude of the Coriolis force [2] experienced by the structure moving with a mean linear velocity v and angular velocity Ω can be expressed as

$$F_{Coriolis} = -2\rho\left|\left(\vec{\Omega}\times\vec{v}\right)\right| = -2\rho\Omega v\sin\theta, \tag{7}$$

where, θ is the angle between $\vec{\Omega}$ and $\vec{v}$, here being considered as constant. Now, we apply a crude simplification approximation

($v \sim P^{1/2}\rho^{-1/2}$) for analytic convenience from the kinetic theory of fluid [5]. For the description of the bulk macroscopic state of the stellar matter, eq. (7) can now be given as

$$F_{Coriolis} = -2\rho\Omega\left(\frac{8P}{\pi\rho}\right)^{1/2}\sin\theta, \tag{8}$$

where, P denotes the isothermal fluid pressure. Lastly, the viscous drag force arising because of internal atomic or molecular friction at a collective level can be obtained by applying the Navier-Stokes equation for the net force density conservation [8]. Thus, the explicit expression for the viscous force counterpart can finally be obtained as

$$F_{viscous} = \mu\left[-\sqrt{\frac{2P}{\pi\rho^3}}\cdot\frac{d^2\rho}{dr^2}+\sqrt{\frac{2}{P\pi\rho}}\cdot\frac{d^2P}{dr^2}+\sqrt{\frac{9P}{2\pi\rho^5}}\cdot\left(\frac{d\rho}{dr}\right)^2 -\sqrt{\frac{1}{2\pi\rho P^3}}\cdot\left(\frac{dP}{dr}\right)^2-\sqrt{\frac{2}{P\pi\rho^3}}\left(\frac{dP}{dr}\right)\left(\frac{d\rho}{dr}\right)\right], \tag{9}$$

where, $\mu = \rho\nu$ is the absolute viscosity with ν as the kinematic viscosity, also known as the momentum diffusivity [3].

Now, applying the basic concept of radiation hydrodynamics [11-12] in mind, we have the radiation force as

$$F_{rad} = \frac{g}{\alpha}\left(\frac{1}{3}aT^4\right),$$

which can further be simplified as,

$$F_{rad} = \frac{g}{\alpha}\left[\frac{1}{3}a\left(\frac{PR}{M\rho}\right)^4\right]. \tag{10}$$

where, a is the (thermal) radiation constant and R is the universal (molar) gas constant [1-3].

Applying all the derived expressions for the force laws from the above equations (eqs. (5)-(6) and eqs. (8)-(10)), and simplifying eq. (1) with $\rho = P/\alpha$, such that $\alpha \sim \omega gr \approx \omega g\lambda_J$ in the customary notations as a new gravitational coupling parameter, one finally obtains a net pressure balance equation in in the proposed formalism as

$$X\left(\frac{dP}{dr}\right)^2 - Y\frac{P}{r}\left(\frac{dP}{dr}\right) + ZP^3 + \frac{P^3}{r} - A^*P^4r^2 = 0, \tag{11}$$

where, $X = \mu\left(\sqrt{9\alpha/2\pi} - \sqrt{\alpha/2\pi} - \sqrt{2\alpha/\pi}\right)$, $Y = GAm^2/5r$, $Z = g/\alpha + \sqrt{32/\alpha\pi}\Omega\sin\theta$, and $A^* = 8\pi/m\alpha + g\alpha^3R^4/M^4$.

The focal goal of our analysis lies in characterizing the considered stellar structure configuration in a standard scale-free invariant form. Accordingly, we transform all the relevant physical parameters (dimensional) in a normalized (dimensionless) form by using a standard astrophysical normalization scheme [9]. The new set of parameters is described as $\xi := r/\lambda_J$ for the normalized distance and $P^* := P/P_0$ for the normalized pressure. Here, $\lambda_J = (kT/4\pi Gm\rho)^{1/2}$ is the well-known Jeans scale length. Therefore, eq. (11) with these reservations in the normalized form can be written as

$$X\left(\frac{dP^*}{d\xi}\right)^2 - Y\frac{P^*}{\xi}\frac{dP^*}{d\xi} + Z\beta(P^*)^3 + \gamma\frac{(P^*)^3}{r} - A^*\beta^2(P^*)^4\xi^2 = 0, \tag{12}$$

where, $\beta = P_0\lambda_J^{\ 2}$ and $\gamma = P_0\lambda_J$.

After a little bit of algebraic simplification, eq. (12) describing the equilibri um barotropic dynamics on the Jeans scale of space ($r \sim \lambda_J$) reduces to a new form as

$$\left(\frac{dP^*}{d\xi}\right)^2 - a_1\left(\frac{1}{\sqrt{T}}\right)\frac{P^*}{\xi}\frac{dP^*}{d\xi} + \left(a_2\sqrt{T} - a_3 T\right)\left(P^*\right)^3$$
$$+ \left(a_4 T\sqrt{\rho}\right)\frac{\left(P^*\right)^3}{\xi} - a_5\sqrt{T^5}\left(P^*\right)^4\xi^2 = 0, \qquad (13)$$

where, the different coefficients actively involved in eq. (13) can respectively be presented as

$$a_1 = \frac{Gm^{5/2}A}{\mu\left(\sqrt{9gk\lambda_J} - \sqrt{gk\lambda_J} - \sqrt{4gk\lambda_J}\right)}, \qquad (14)$$

$$a_2 = \frac{k}{4\pi mG\lambda_J\mu\left(\sqrt{9gk\lambda_J} - \sqrt{gk\lambda_J} - \sqrt{4gk\lambda_J}\right)}, \qquad (15)$$

$$a_3 = \frac{\sqrt{32}\Omega k^{3/2}\sin\theta}{4\pi^{3/2}Gm^{3/2}g^{1/2}\lambda_J^{1/2}\mu\left(\sqrt{9gk\lambda_J} - \sqrt{gk\lambda_J} - \sqrt{4gk\lambda_J}\right)}, \qquad (16)$$

$$a_4 = \frac{k^{3/2}}{2m^{3/2}\pi^{1/2}G^{1/2}\mu\left(\sqrt{9gk\lambda_J} - \sqrt{gk\lambda_J} - \sqrt{4gk\lambda_J}\right)}, \qquad (17)$$

$$a_5 = \frac{k^3 + g^4\lambda_J^3R^4}{2\pi\mu m^4G^2M^4\left(\sqrt{9gk\lambda_J} - \sqrt{gk\lambda_J} - \sqrt{4gk\lambda_J}\right)}. \qquad (18)$$

3. RESULTS AND DISCUSSIONS

A new radial barotropic equation of state (eq. (13)) for theoretically characterizing the equilibrium structure of complex stellar configurations in a normalized (scale-free) form with spherical symmetry is analytically developed. An exploration for the explicit analytical solutions, even in an approximate form, is found to be tedious. A numerical analysis for investigating the exact bounded solutions with the help of the fourth–order Runge-Kutta method [10] is constructed. In order to characterize the average dynamical features, we choose the quantitative input values of the different multi-parametric coefficients in eq. (13) in a crude approximate form as $a_1 \approx a_2 \approx a_3 \approx a_4 \sim 10^{-4}$ and $a_5 \approx 10^{-2}$. The initial seed value of the pressure is taken as $P_0^* = 10^{-3}$. The numerical results, thus obtained by the above integration technique, are graphically displayed in Figures 1-2.

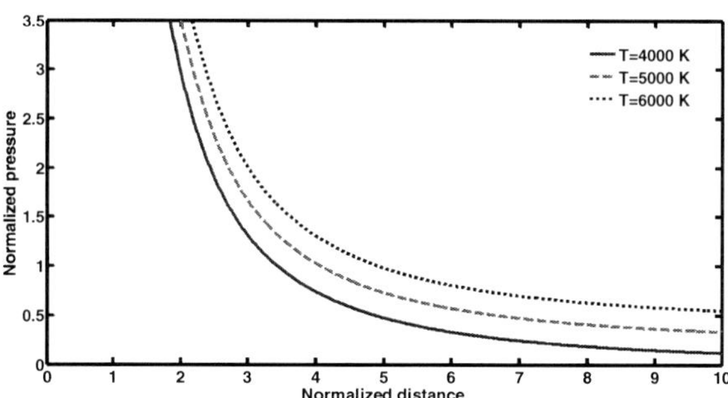

Figure 1. Spatial profile of the normalised pressure (P^*) for different values of temperature (T), but with a fixed densiy (ρ =1400 kg m-3). The various lines refer to T=4000 K (blue solid line), T=5000 K (red dashed line), and T=6000 K (black dotted line); respectively. The fine details are presented in the text.

In Figure 1, we portray the spatial variation of the normalized stellar pressure as a hydrostatically bounded configuration for three different values of the stellar temperature, but with a fixed densiy value, ρ =1400 kg m^{-3}. The various lines correspond to T=4000 K (blue solid line), T=5000 K (red dashed line), and T=6000 K (black dotted line); respectively. It is seen that the effective normalized pressure attains a maximum value at the centre of the stellar fluid mass distribution. It then asymptotically decreases to some non-zero finite value from the centre outwards. It is further found that, more the stellar temperature, more is the localised pressure; and vice-versa. Similar features, as Figure 1, are found to exist even in Figure 2, but for different density values with a fixed temperature, T=5000 K. The various lines now correspond to ρ=1400 kg m^{-3} (blue solid line), ρ=1600 kg m^{-3} (red dashed line), and ρ=1800 kg m^{-3}(black dotted line); respectively. It is speculated that the effective localised normalised pressure increases with the material density, and vice-versa. In both the cases (Figures 1-2), the maximization of the effective pressure is attributable to self-gravitational condensation for bounded stellar structures to form and evolve. This is indeed worth mentioning here that the enhanced value of the effective thermal pressure at the centre acts as a unique agent to trigger thermonuclear reactions thereby yielding a tremendous amount of energy during the star formation processes. The profile patterns are interestingly found to be in good conformity with the previously reported results stemming from the polytropic stellar configurations (see Figures 1-3 in [5] and Figure 2 in [9]).

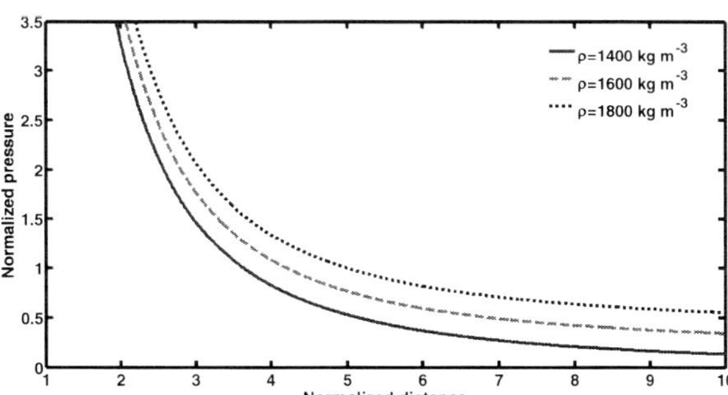

Figure 2. Same as Figure 1, but for different densities with fixed temperature (T=5000 K). The various lines link to ρ =1400 kg m-3 (blue solid line), ρ =1600 kg m-3 (red dashed line), and ρ =1800 kg m-3 (black dotted line); respectively.

CONCLUSIONS

In conclusive summary, a new model equation of macroscopic state to investigate the gyro-equilibrium dynamical properties of complex stellar structures in the presence of various unavoidable realistic parametric factors with respect to a statical inertial frame of reference is theoritically devised. The adopted classical model considers the conjoint balanced action of all the multi-parametric fluid-functional agencies, such as viscous force (inward), gravitational force (inward), fluid turbulence force (outward), force due to the Coriolis rotation (outward), and force due to radiation hydrodynamics (outward) relative to the centre of the stellar material mass distribution concurrently. The resulting new dimensional radial equation of state is transformed into a standard normalized (dimensionless)form for a scale-free description of the stellar equilibrium characteristic features. Analytical complications involved in the mathematical construct of the methodoligically derived equation opens a new challenge for finding explicit bounded solutions enabling structure formation from a new viewpoint of analytic integrability perspective. We, therefore, construct a detailed numerical illustrative scheme to obtain the exact solutions of the derived barotropic equation (eq. (13)) via the fourth-order Runge-Kutta method [10]. Our classical non-relativistic analysis establishes a direct correlation between the net stellar pressure in the parametric space defined by the stellar temperature and material density; and so forth. Moreover, we see that pressure increases to a large extent when density and temperature increase. It can be related to the fact that as the star reaches its end it becomes denser and denser. As a result, the density increases. Since, the stellar pressure is directly proportional to density, so with increase in density, the pressure increases. And the same is true even for the temperature variation. The practical reliability of the investigated semi-analytic results as special corollaries and isolated cases are piecewise examined and contextually validated in the panoptic light of the previous conformal predictions for bounded polytropic stellar configurations reported elsewhere [1-2]. In addition, our numerical calculation scheme is in good agreement with the relevant thermo-dynamical profiles for self-gravitationally evolving diversified stellar structures as extensively evident from various reliable sources, such as [1], [15], etc.

ACKNOWLEDGMENTS

Active cooperation received from Tezpur University is thankfully acknowledged. We are likewise grateful to the Editorial Board Members. The financial support received (by the corresponding author) through the SERB Project (Grant- EMR / 2017 / 003222) is duly recognized.

REFERENCES

[1] Chandeasekhar, S., (2016). *An Introduction to the Study of Stellar Structure* (Dover, USA).

[2] Kippenhahn, R., Weigert, A. & Weiss, A., (2012). *Stellar Structureand Evolution* (Springer, Berlin).

[3] Franco, J. & Carraminana, A. (eds.), (1999). *Interstellar Turbulence* (Cambridge Univ. Press, UK).

[4] Borah, B. & Karmakar, P. K., (2015). "A theoretical model for electromagnetic characterization of a spherical dust molecular cloud equilibrium structure," *New Astron.* 40:49.

[5] Hunter, C., (2001). "Series solutions for polytropes and the isothermal sphere," *Mon. Not. Royal Astron. Soc.* 328:839.

[6] Chandrasekhar, S., (1951). "The gravitational instability of an infinite homogeneous turbulent medium," *Proc. Royal Soc.* London. Ser. A 210:26.

[7] Chandrasekhar, S. & Fermi, E., (1953). "Problems of gravitational stability in the presence of a magnetic field," *Astrophys. J.* 118:116.

[8] Vazquez-Semadeni, E., Canto, J. & Lizano, S., (1998). "Does turbulent pressure behave as a logatrope?," *Astrophys. J.* 492:596.

[9] Maeder, A., (2009). *Physics, Formation, and Evolution of Rotating Stars* (Springer, Heidelberg).

[10] Lindfield G. R & Penny, J. E. T., (2012). *Numerical Methods Using MATLAB* (Elsevier, UK).

[11] Braiding, C. R., Wardle, M., (2012). "The Hall effect in accretion flows," *Mon. Not. Royal Astron. Soc.* 427:3188.

[12] Castor, J. I., (2004). *Radiation Hydrodynamics* (Cambridge Univ. Press, UK).

[13] Miller, J. L., (2020). "Compressibility measurements reach white dwarf pressures," *Physics Today* 73:14.

[14] Das, P. & Karmakar, P. K., (2017). "Instability behaviour of cosmic gravito-coupled correlative complex bi-fluidic admixture," *Europhys. Lett.* 120:19001.

[15] Cadez, V. M., (1990). "Applicability problem of Jeans criterion to a stationary self-gravitating cloud," *Astron. Astrophys.* 235:242.

In: An Introduction to Molecular Clouds ISBN: 978-1-53619-178-3
Editor: Sachin Kaothekar

Chapter 4

JEANS INSTABILITY OF ROTATING PLASMA WITH RADIATIVE HEAT-LOSS FUNCTION AND FLR CORRECTIONS FLOWING THROUGH POROUS MEDIUM FOR MOLECULAR CLOUD CONFIGURATION

Sachin Kaothekar*

Department of Physics, Mahakal Institute of Technology & Managemant, Ujjain (M.P.) - 456664, India

ABSTRACT

The effect of rotation, porosity and finite ion Larmor radius (FLR) corrections on the gravitational instability and radiative instability of infinite homogeneous plasma has been explored integrating the consequences of radiative heat-loss function and thermal conductivity. The general dispersion relation is obtained by means of the normal mode

* Corresponding Author's Email: sackaothekar@gmail.com, sachinmgi007@gmail.com.

analysis scheme with the help of appropriate linearized perturbation equations of the predicament. This dispersion relations is further condenses for rotation axis parallel and perpendicular to the magnetic field. Stability of the medium is argued by pertaining Routh Hurwitz's criterion and it is found that Jeans criterion establishes the stability of the medium. We locate that the presence of radiative heat-loss function and thermal conductivity amend the fundamental Jeans criterion of gravitational instability into radiative instability criterion. Numerical computations have been executed to show the effect of various parameters on the growth rate of the Jeans-gravitational instability. We find that rotation, FLR corrections and medium porosity steady the growth rate of the organization in the transverse mode of propagation. Our result demonstrates that the rotation, porosity and FLR corrections affect the dens molecular clouds configuration and star formation.

Keywords: jeans instability, rotation, FLR correction, radiative heat-loss functions, flow through porous media

1. INTRODUCTION

In the recent era of research for the field of physics, plasma instabilities is studied to understand the process of formation of small and big structures in astronomy and astrophysical. Now a day's plasma physics have been one of the most important areas of research in physics. It is a well established fact that Jeans-gravitational instability and rotation plays a key role in understanding the process of formation of molecular clouds, stars, planets, asteroids, comets and other astrophysical objects. In this direction, a detailed account of the Jeans-gravitational instability with rotation and magnetic field has been given by Chandrasekhar [1]. In this connection many authors (Houser [2], Soni and Chhajlani [3], Tandon and Talwar [4], Prajapati et al. [5] and Uberoi [6]) have discussed the problem of Jeans gravitational instability of plasma taking the effect of rotation. Vyas and Chhajlani [7] have investigated the problem of gravitational instability with rotation, magnetic field, viscosity, and thermal conductivity. Khan and Bhatia

[8] have carried out the effects of rotation, Hall current and finite electrical conductivity on gravitational instability of plasma. Prajapati et al. [9] have studied the influence of rotation on Jeans-gravitational instability of anisotropic heat-conducting plasma and found that rotation has stabilizing influence on the growth rate of the system. Thus we find that rotation plays an important role in the investigation of the problem of gravitational instability.

In addition to this, the difficulty of gravitational instability of plasma flowing through porous medium has much importance in the study of large and small astrophysical objects, such as comets, meteorites and interplanetary dust. More over the study of flow through porous media is of considerable interest due to its variety of applications in geophysical situations, magneto -hydrodynamics (MHD) flows, laboratories, industries and in petroleum and chemical engineering. Much of the pioneer work in the field of plasma floe through porous medium is reviewed by Nield and Bejan [10] and Vafai [11]. Vyas and Chhajlani [12] have discussed the magnetogravitational instability of an infinite homogeneous rotating plasma through a porous medium with finite electrical and thermal conductivities. Sharma and Thakur [13] have investigated the gravitational instability of flow through porous medium for some system of astrophysical interest. The effect of porosity on gravitational instability is also investigated by several authors (Kaothekar and Chhajlani [14], El-Sayed and Mohamed [15], Rana [16], and El-Sayed and Hussein [17]), and show that porosity of the medium has a stabilizing influence on the growth rate of the medium. Chand and Rana [18] have investigated the effect of radiation on the onset of thermal instability in a layer of nanofluid layer in a porous medium. Recently Chand et al. [19] have discussed the thermal instability in Rivlin-Ericksen elasto viscous nanofluid in a porous medium: a revised model. Recently Kaothekar and Chhajlani [20] have investigated Jeans instability of self gravitating partially ionized Hall plasma with radiative heat loss functions and porosity. Thus we see that, porosity of the medium plays a crucial role in instability and stability examinations of the gravitating magnetized plasma flowing through porous medium.

Along with this, in recent years the importance of FLR in gravitational instability of plasma is emphasized due to its large applications in astrophysics. Many investigators (Jukes [21], Roberts and Taylor [22], Rosenbluth et al. [23], and Singh and Hans [24]) have examined the stabilizing influence of finite ion Larmor radius (FLR) which exhibits itself in the form of magnetic viscosity in the fluid dynamic equations, on the plasma instability. Herrnegger [25] in his study of effects of collision and gyroviscosity on gravitational instability in a two-component plasma shows that the critical wave number decreases with increasing the FLR. Sharma [26] has investigated the stabilizing role of FLR with rotation and magnetic field on gravitational instability of plasma. Chhonkar and Bhatia [27] have carried out the effect of FLR on the gravitational instability of two-component plasma with viscosity magnetic resistivity and Hall current. Recently Devlen and Pekunlu [28] have investigated the effects of FLR on weakly magnetized, dilute plasmas. More recently Kaothekar and Chhajlani [29] have studied the effect of FLR corrections on Jeans instability of viscous thermally conducting Jeans gravitating astrophysical plasma. Thus we see that FLR is an important parameter in discussion of Jeans-gravitational instability and other hydrodynamic instability.

In addition to this, the recent observations from solar corona, stellar regions and interstellar medium have shown that the radiative heat-loss mechanism plays an important role in the process of molecular clouds condensation and in star formation, in connection with thermal instability. This radiative heat-loss function represents the decay of heat in system with respect to local temperature and density. Several authors (Field [30], Hunter [31], Iabnez [32], and Talwar and Bora [33]) have investigated the phenomenon of thermal instability arising due to heat-loss mechanism in plasmas and discussed its role in the formation of solar prominences, condensation in planetary nebula and in the condensation of galaxies from the intergalactic medium. In this direction, Burket and Lin [34] have investigated the thermal instability in the formation of clumpy gas clouds and they showed that the thermal instability can lead to the breakup of large clouds into cold dense clumps. Radwan [35] has investigated the role of radiative heat-loss function in the study of self-gravitational instability of radiating rotating gas cloud streams with non-uniform velocity. Aggarwal and Talwar [36] have pointed out the role of radiative heat-loss function in the study of magnetothermal instability in a rotating gravitating fluid. Bora and Talwar [37] have investigated the problem of magneto-thermal instability with radiative heat-loss function, Hall current, finite resistivity, electron inertia and thermal conductivity. Prajapati et al. [38] have investigated the self-gravitational instability of rotating viscous Hall plasma with arbitrary radiative heat-loss function and electron inertia. Kaothekar et al. [39] have carried out the effect of radiative heat-loss function and neutral collisions on self gravitational instability of viscous thermally conducting partially-ionized plasma. Sharma and Jain [40] have investigated Radiative condensation instability in partially ionized dusty plasma with polarization force. Prajapati et al. [41] have studied Jeans instability in a coalitional strongly coupled dusty plasma with radiative condensation with polarization force. Recently Kaothekar [42] has explored the problem of thermal instability of partially ionized viscous plasma with Hall effect FLR corrections flowing through porous medium. More recently Kaothekar [43] has investigated the Jeans instability of rotating plasma with radiative heat-loss function and FLR corrections flowing through porous medium. Thus it is clear that radiative heat-loss function is an important parameter for discussion of Jeans gravitational instability.

From the above studies we find that joint influence of the rotation, FLR corrections, radiative heat-loss function, thermal conductivity and on the gravitational instability is not studied. Thus keeping in mind the importance of rotation and FLR corrections in formation of astrophysical small and big objects, we endeavor to discuss the effect of rotation, porosity and FLR corrections on Jeans-gravitational instability of plasma with thermal conductivity and radiative heat-loss function.

2. Basic Equations of the Problem

We consider an infinite homogeneous, self-gravitating, thermally conducting, radiating, porous plasma of with FLR corrections in the presence of magnetic field *H* (0, 0, *H*). The equations of the problem with these effects are written as-

$$\frac{1}{\varepsilon}\frac{d\boldsymbol{u}}{dt} = -\frac{\nabla p}{\rho} - \frac{\nabla \cdot \mathbf{P}}{\rho} + \nabla \boldsymbol{U} + 2(\boldsymbol{u} \times \boldsymbol{\Omega}) + \frac{1}{4\pi\rho}(\nabla \times \boldsymbol{H}) \times \boldsymbol{H}, \quad (1)$$

$$\varepsilon \frac{d\rho}{dt} + \rho \nabla . \boldsymbol{u} = 0 , \quad (2)$$

$$\frac{1}{\gamma - 1}\frac{dp}{dt} - \frac{\gamma}{\gamma - 1}\frac{p}{\rho}\frac{d\rho}{dt} + \rho L - \nabla.(\lambda \nabla T) = 0, \quad (3)$$

$$p = r\, RT , \quad (4)$$

$$\frac{\partial \boldsymbol{B}}{\partial \mathrm{t}} = \frac{1}{\varepsilon}\nabla \times (\boldsymbol{u} \times \boldsymbol{H}), \quad (5)$$

$$\nabla . \boldsymbol{H} = 0, \quad (6)$$

$$\nabla^2 \boldsymbol{U} + 4\pi G \rho = 0, \quad (7)$$

where p, ρ, υ, T, $\boldsymbol{u}$ (u_x, u_y, u_z), $\boldsymbol{\Omega}(W_x, 0, W_z)$, l, e, $\boldsymbol{U}$, G, R, and g denote the fluid pressure, density, kinematic viscosity, temperature, velocity, rotation, thermal conductivity, porosity, gravitational potential, gravitational constant, gas constant and ratio of two specific heats respectively. L (ρ, T) is the heat-loss function of the material exclusive of thermal conduction and is in general a function of the local values of density and temperature. The operator (d/dt) is the substantial derivative given as $(d/dt) = (\partial/\partial t + (1/\varepsilon)\, \boldsymbol{u}.\nabla)$.

3. Linearized Perturbation Equations

The perturbation in fluid pressure, density, temperature, velocity, magnetic field, heat-loss function and gravitational potential are given as δp, $\delta\rho$, δT, u (δu_x δu_y, δu_z), $\boldsymbol{\delta H}$ (δH_x, δH_y, δH_z), L and δU respectively. The perturbation state is given as

$$p = p_0 + \delta p,\ \rho = \rho_0 + \delta\rho,\ T = T_0 + \delta T,\ u = u_0 + \boldsymbol{\delta u} \tag{8}$$

$$\boldsymbol{H} = \boldsymbol{H}_0 + \boldsymbol{\delta H},\ L = L_0 + L \text{ and } \boldsymbol{U} = \boldsymbol{U}_0 + \boldsymbol{\delta U}.$$

Suffix '0' represents the initial equilibrium state, which is independent of space and time.

Substituting the perturbation state into equation (1) to (7) and linearizing them by neglecting higher order perturbations. Suffix '0' is dropped from the equilibrium quantities.

The linearized perturbation equations of motion for such medium are

$$\frac{1}{\varepsilon}\frac{\partial \boldsymbol{\delta u}}{\partial t} = -\frac{\nabla \delta p}{\rho} - \frac{\nabla \cdot \mathbf{P}}{\rho} + \nabla \delta U + 2(\boldsymbol{u} \times \boldsymbol{\Omega}) + \frac{1}{4\pi\rho}(\nabla \times \boldsymbol{\delta H}) \times \boldsymbol{H}, \tag{9}$$

$$\varepsilon \frac{\partial \delta \rho}{\partial t} + \rho \nabla . \delta \boldsymbol{u} = \mathbf{0}, \tag{10}$$

$$\frac{1}{\gamma - 1} \frac{\partial \delta p}{\partial t} - \frac{\gamma}{\gamma - 1} \frac{p}{\rho} \frac{\partial \delta \rho}{\partial t} + \rho \left[\delta \rho \left(\frac{\partial L}{\partial \rho} \right)_T + \delta T \left(\frac{\partial L}{\partial T} \right)_\rho \right] - \lambda \nabla^2 \delta T = 0, \tag{11}$$

$$\frac{dp}{p} = \frac{dT}{T} + \frac{dr}{r}, \tag{12}$$

$$\frac{\partial \delta \boldsymbol{H}}{\partial \mathrm{t}} = \frac{1}{\varepsilon} \nabla \times (\boldsymbol{u} \times \boldsymbol{H}), \tag{13}$$

$$\nabla . \delta \boldsymbol{H} = 0, \tag{14}$$

$$\nabla^2 \delta \boldsymbol{U} + 4\pi G \delta \rho = 0, \tag{15}$$

where $\left(\partial L/\partial T\right)_\rho$, $\left(\partial L/\partial \rho\right)_T$ are the partial derivatives of temperature dependent heat-loss function L_T and density dependent heat-loss function L_r respectively. The components of pressure tensor P, considering the finite ion gyration radius for the magnetic field along z-axis as given by Roberts and Taylor (1962) are

$$P_{xx} = -\rho \upsilon_0 \left(\frac{\partial \delta u_y}{\partial x} + \frac{\partial \delta u_x}{\partial y} \right), \; P_{yy} = \rho \upsilon_0 \left(\frac{\partial \delta u_y}{\partial x} + \frac{\partial \delta u_x}{\partial y} \right),$$

$$P_{xy} = P_{yx} = \rho \upsilon_0 \left(\frac{\partial \delta u_x}{\partial x} - \frac{\partial \delta u_y}{\partial y} \right), \; P_{xz} = P_{zx} = -2\rho \upsilon_0 \left(\frac{\partial \delta \mathrm{v}_y}{\partial z} + \frac{\partial \delta \mathrm{v}_z}{\partial y} \right),$$

$$P_{yz} = P_{zy} = 2\rho \upsilon_0 \left(\frac{\partial \delta u_z}{\partial x} + \frac{\partial \delta u_x}{\partial z} \right), \; P_{zz} = 0. \tag{16}$$

The parameter u_0 has the dimensions of the kinematics viscosity and called as magnetic viscosity defined as $u_0 = W_L R_L^2/4$, where R_L is the ion-Larmor radius and W_L is the ion gyration frequency.

We seek plain wave solution of the form

$$\exp\left(is\,t + ik_x x + ik_z z\right), \tag{17}$$

where s is the frequency of harmonic disturbance, k_x and k_z are the wave numbers of the perturbations along x and z axes.

The components of equation (14) may be given

$$\delta H_x = \frac{iH}{\varepsilon\omega} k_z \delta u_x, \quad \delta H_y = \frac{iH}{\varepsilon\omega} k_z \delta u_y, \quad \delta H_z = -\frac{iH}{\varepsilon\omega} k_x \delta u_x. \tag{18}$$

where $is = w$

Using equations (12), (13) and (17) we write

$$\delta p = \frac{\left\{(\gamma-1)\left[TL_T - \rho L_\rho + (\lambda k^2 T/\rho)\right] + \omega c^2\right\}}{\left\{(\gamma-1)\left[(T\rho/p)L_T + (\lambda k^2 T/p)\right] + \omega\right\}}\delta\rho, \tag{19}$$

Using equations (11)-(19) in equation (10), we may write the following algebraic equations for the components of equation (10)

$$\delta u_x\left[\omega + \frac{V^2k^2}{\omega}\right] + \delta u_y\left[\varepsilon\upsilon_0(k_x^2 + 2k_z^2) - 2\varepsilon\Omega_z\right] + \varepsilon\frac{ik_x}{k^2}\Omega_T^2\,\mathrm{s} = 0, \tag{20}$$

$$-\delta u_x\left[\varepsilon\upsilon_0(k_x^2 + 2k_z^2) - 2\Omega_z\right] + \delta u_y\left[\omega + \frac{V^2k_z^2}{\omega}\right] - \delta u_z\left[2\varepsilon\left(\upsilon_0 k_x k_z + \Omega_x\right)\right] = 0, \tag{21}$$

$$\delta u_y\left[2\varepsilon\left(\upsilon_0 k_x k_z + \Omega_x\right)\right] + \delta u_z\omega + \varepsilon\frac{ik_z}{k^2}\Omega_T^2\,\mathrm{s} = 0. \tag{22}$$

Taking divergence of equation (10) and using equation (11) to (19), we obtain as

$$\delta u_x\left[ik_x\frac{V^2k^2}{\varepsilon\omega}\right]+\delta u_y\left[i\upsilon_0k_x\left(k_x^2+4k_z^2\right)+2i\left(k_z\Omega_x-k_x\Omega_z\right)\right]-\mathrm{s}\left[\omega^2+\Omega_T^2\right]=0, \qquad (23)$$

we have made following substitutions

$$\alpha=(\gamma-1)\left(TL_T-\rho L_\rho+\frac{\lambda k^2T}{\rho}\right),\ \beta=(\gamma-1)\left(\frac{T\rho L_T}{p}+\frac{\lambda k^2T}{p}\right),\ s=\delta\rho/\rho,$$

$$\Omega_T^2=\frac{\Omega_I^2+\omega\Omega_J^2}{\omega+\beta},\ \Omega_J^2=c^2k^2-4\pi G\rho,\ \Omega_I^2=k^2\alpha-4\pi G\rho\beta,\ V^2=\frac{H^2}{4\pi\rho},$$

$c=(gp/r)^{1/2}$ is the adiabatic velocity of sound in the medium. (24)

4. Dispersion Relation

The nontrivial solution of the determinant of the matrix obtained from equations (20)-(23) with $\delta\mathrm{v}_x, \delta\mathrm{v}_y, \delta\mathrm{v}_z, s$ having various coefficients that should vanish is to give the following dispersion relation

$$\left(\omega^2+\Omega_T^2\right)\left[\omega+\frac{V^2k^2}{\omega}\right]\left\{\omega\left[\omega+\frac{V^2k_z^2}{\omega}\right]+4\varepsilon^2\left(\upsilon_0k_xk_z+\Omega_x\right)^2\right\}-\frac{2\varepsilon^2\Omega_T^2}{k^2}\left(\upsilon_0k_xk_z+\Omega_x\right)$$
$$\times\left[\omega+\frac{V^2k^2}{\omega}\right]\left[\upsilon_0k_xk_z\left(k_x^2+4k_z^2\right)+2\left(k_z^2\Omega_x-k_xk_z\Omega_z\right)\right]+\omega\left[\varepsilon\upsilon_0\left(k_x^2+2k_z^2\right)-2\varepsilon\Omega_z\right]^2$$
$$\times\left(\omega^2+\Omega_T^2\right)+\frac{2\varepsilon k_xk_zV^2}{\omega}\Omega_T^2\left(\upsilon_0k_xk_z+\Omega_x\right)\left[\varepsilon\upsilon_0\left(k_x^2+2k_z^2\right)-2\varepsilon\Omega_z\right]-\omega\frac{\varepsilon\Omega_T^2}{k^2}\left[\varepsilon\upsilon_0\right.$$
$$\left.\times\left(k_x^2+2k_z^2\right)-2\varepsilon\Omega_z\right]\left[\upsilon_0k_x^2\left(k_x^2+4k_z^2\right)+2\left(k_xk_z\Omega_x-k_x^2\Omega_z\right)\right]-\omega\left[\omega+\frac{V^2k_z^2}{\omega}\right]$$
$$\times\frac{V^2k_x^2}{\omega}\Omega_T^2-\frac{4\varepsilon^2k_x^2V^2}{\omega}\Omega_T^2\left(\upsilon_0k_xk_z+\Omega_x\right)^2=0. \qquad (25)$$

The dispersion relation (25) shows the combined influence of rotation, FLR corrections, radiative heat-loss function and thermal conductivity on the Jeans-gravitational instability of a homogeneous plasma flowing through porous medium. The dispersion relation (25) may be easily verified with the previous known results. In absence of radiative heat-loss function, thermal conductivity and the general dispersion relation (25) is identical to Sharma [26]. In absence of FLR corrections the general dispersion relation (25) reduces to Prajapati et al. [41] neglecting Hall current, electron inertia and viscosity in their case. If we neglect the effect of rotation the general dispersion relation (25) is identical to Kaothekar and Chhajlani [20] neglecting viscosity in their case. The above dispersion relation is lengthy and to study the effect of each parameter we now reduce the dispersion relation (25) for transverse and longitudinal modes of propagation.

4.1. Discussion of the Dispersion Relation Transverse Mode of Propagation (K ⊥ B)

In this case the perturbations are taken to be perpendicular to the direction of the magnetic field $(i.e.,\ k_x = k,\ k_z = 0)$. The dispersion relation (25) reduces to

$$\left(\omega^2+\Omega_T^2\right)\left\{\left[\omega+\frac{V^2k^2}{\omega}\right]\left(\omega^2+4\varepsilon^2\Omega_x^2\right)+\omega\left(\varepsilon\upsilon_0k^2-2\varepsilon\Omega_z\right)^2\right\}-\Omega_T^2\left\{\omega\left[\frac{\omega V^2k^2}{\omega}+\varepsilon^2\left(\upsilon_0k^2-2\Omega_z\right)^2\right]+4\varepsilon^2\Omega_x^2\frac{V^2k^2}{\omega}\right\}=0. \quad (26)$$

The dispersion relation (26) gives the influence of rotation, FLR corrections, radiative heat-loss function thermal conductivity and porosity on Jeans-gravitational instability of plasma for transverse mode of propagation. Now we discuss the dispersion relation (26) for rotation axis parallel and perpendicular to the magnetic field.

4.1.1. Axis of Rotation along the Magnetic Field ($\Omega||B$)

For the case of axis of rotation along the magnetic field, we substitute $\Omega_x = 0$ and $\Omega_z = \Omega$ in dispersion relation (26) which reduces to

$$\omega^3\left\{\omega\left[\omega+\frac{V^2k^2}{\omega}\right]+\varepsilon^2\left(\upsilon_0k^2-2\Omega\right)^2+\frac{\Omega_I^2+\omega\Omega_J^2}{\omega+\beta}\right\}=0. \tag{27}$$

In absence of FLR corrections this dispersion relation (27) is identical to that obtained by Prajapati et al. [41] neglecting viscosity and electron inertia in their case. Equation (27) has two independent factors. The first factor of equation (27) gives $\omega^3 = 0$, which is a marginal stable mode. The second factor of equation (27) gives the following dispersion relation on substituting the values of $\Omega_I^2, \Omega_J^2, \alpha$ and β

$$\omega^3+\left\{\left[(\gamma-1)\left(\frac{T\rho L_T}{p}+\frac{\lambda k^2T}{p}\right)\right]\right\}\omega^2+\left[4\varepsilon^2\Omega^2+\varepsilon^2\upsilon_0^2k^4+V^2k^2+c^2k^2-4\pi G\rho\right.$$
$$\left.-4\varepsilon^2\Omega\upsilon_0k^2\right]\omega+\left[(\gamma-1)\left(\frac{T\rho L_T}{p}+\frac{\lambda k^2T}{p}\right)\right]\left(4\varepsilon^2\Omega^2+\varepsilon^2\upsilon_0^2k^4+V^2k^2-4\varepsilon^2\Omega\upsilon_0k^2\right)$$
$$+k^2(\gamma-1)\left(TL_T-\rho L_\rho+\frac{\lambda k^2T}{\rho}\right)-4\pi G\rho(\gamma-1)\left(\frac{T\rho L_T}{p}+\frac{\lambda k^2T}{p}\right)\Bigg\}=0. \tag{28}$$

This dispersion relation represents the effect of simultaneous inclusion of rotation, FLR corrections, radiative heat-loss function, thermal conductivity and porosity on the Jeans-gravitational instability of the system. In absence of FLR corrections and rotation dispersion relation (28) is identical to Bora and Talwar [37] neglecting electron inertia in their case. On neglecting the effect of FLR corrections dispersion relation (28) reduces to that of Aggarwal and Talwar [36]. In absence of FLR corrections dispersion relation (28) is identical to

Prajapati et al. [41] for $(f = 1, u = 0)$ in their case. When constant term of equation (28) is less than zero this allows at least one positive real root which corresponds to the instability of the system. The condition of instability obtained from constant term of equation (28) is given as

$$\left\{k^2\left(TL_T - \rho L_\rho + \frac{\lambda k^2 T}{\rho}\right) - \left(\frac{T\rho L_T}{p} + \frac{\lambda k^2 T}{p}\right)\left(4\varepsilon^2\Omega^2 + \varepsilon^2 \upsilon_0^2 k^4 + V^2 k^2 - 4\varepsilon^2 \Omega \upsilon_0 k^2\right)\right\} < 0, \qquad (29)$$

Equation (29) represents the modified form of Jeans instability criterion by inclusion of rotation, FLR corrections, radiative heat-loss function and thermal conductivity. In absence of rotation equation (29) is identical to condition (57) of Kaothekar and Chhajlani [20]. If we remove the effect of radiative heat-loss function and FLR corrections in equation (29) then it is identical to equation (45) of Prajapati et al. [41]. So we improve the result of Prajapati et al. [41] by FLR corrections and Kaothekar and Chhajlani [20] by rotation. From equation (29) we conclude that rotation and FLR corrections stabilize the radiative instability.

In absence of FLR corrections $(h = u_0 = 0)$ equation (28) becomes

$$\omega^3 + \left[(\gamma - 1)\left(\frac{T\rho L_T}{p} + \frac{\lambda k^2 T}{p}\right)\right]\omega^2 + \left[4\varepsilon^2 \Omega^2 + V^2 k^2 + c^2 k^2 - 4\pi G\rho\right]\omega + \left\{k^2(\gamma - 1)\right.$$

$$\left.\times\left(TL_T - \rho L_\rho + \frac{\lambda k^2 T}{\rho}\right) - (\gamma - 1)\left(\frac{T\rho L_T}{p} + \frac{\lambda k^2 T}{p}\right)\left[4\pi G\rho - V^2 k^2 - 4\varepsilon^2 \Omega^2\right]\right\} = 0. \qquad (30)$$

On comparing equations (28) and (30) we find that the condition of instability also gets modified. When constant term of equation (30) is less than zero this allows at least one positive real root which corresponds to the instability of the system. The condition of instability obtained from constant term of equation (30) is given as

$$\left\{k^2(\gamma-1)\left(TL_T-\rho L_\rho+\frac{\lambda k^2 T}{\rho}\right)-(\gamma-1)\left(\frac{T\rho L_T}{p}+\frac{\lambda k^2 T}{p}\right)\left[4\pi G\rho-V^2k^2-4\varepsilon^2\Omega^2\right]\right\}<0, \quad (31)$$

The above condition of instability is the modified form of Jeans condition by inclusion of rotation, porosity and magnetic field strength. From equation (34) we conclude that rotation and magnetic field stabilizes the radiative instability. The above inequality (31) is the modified form of Bora and Talwar [36] by inclusion of rotation, porosity and can be solved to yield the following critical wave number.

$$k_{J2}^2=\frac{1}{2}\left|\left[\left(\frac{4\pi G\rho}{c'^2}+\frac{\rho^2 L_\rho}{\lambda T}-\frac{4\varepsilon^2\Omega^2}{c'^2}\right)\left(1+\frac{V^2}{c'^2}\right)^{-1}-\frac{\rho L_T}{\lambda}\right]\pm\left\{\left[\left(\frac{4\pi G\rho}{c'^2}+\frac{\rho^2 L_\rho}{\lambda T}-\frac{4\varepsilon^2\Omega^2}{c'^2}\right)\right.\right.\right.$$

$$\left.\left.\left.\times\left(1+\frac{V^2}{c'^2}\right)^{-1}-\frac{\rho L_T}{\lambda}\right]^2+\frac{16\rho L_T}{\lambda c'^2}\left(\pi G\rho-\varepsilon^2\Omega^2\right)\left(1+\frac{V^2}{c'^2}\right)^{-1}\right\}^{1/2}\right|. \quad (32)$$

In absence of rotation the above modified critical Jeans wave number reduces to that of Aggrawal and Talwar (1969).

In astrophysical situation, instability of the system is one of the most important causes of configuration of objects. So we study the effects of medium porosity ε, rotation Ω^*, and FLR corrections υ_0^* on the growth rate of unstable mode. We present the dispersion relation (28) in a non-dimensional form on dividing it by $(4\pi G\rho)^2$ and the value of γ in numerical calculations is taken as 5/3

$$\omega^{*3}+\left\{\left(L_T^*+\lambda^*k^{*2}\right)\right\}\omega^{*2}+\left[4\varepsilon^2\Omega^{*2}+\varepsilon^2\upsilon_0^{*2}k^{*4}+V^{*2}k^{*2}+k^{*2}-1-4\varepsilon^2\Omega^*\upsilon_0^*k^{*2}\right]\omega$$

$$+\left\{\left[\left(L_T^*+\lambda^*k^{*2}\right)\right]\left(4\varepsilon^2\Omega^{*2}+\varepsilon^2\upsilon_0^{*2}k^{*4}+V^{*2}k^{*2}-4\varepsilon^2\Omega^*\upsilon_0^*k^{*2}\right)\right.$$

$$+k^{*2}\left[\frac{1}{\gamma}\left(L_T^* + \lambda^* k^{*2}\right) - L_\rho^*\right] - \left(L_T^* + \lambda^* k^{*2}\right)\Bigg\} = 0. \qquad (33)$$

where the dimensionless parameters are given as

$$\omega^* = \omega\Big/(4\pi G\rho)^{1/2},\ k^* = ck\Big/(4\pi G\rho)^{1/2},\ L_\rho^* = (\gamma-1)\,\rho L_\rho\Big/c^2(4\pi G\rho)^{1/2},$$
$$L_T^* = (\gamma-1)\,T\rho L_T\Big/p\,(4\pi G\rho)^{1/2},\ \lambda^* = (\gamma-1)\,T\lambda(4\pi G\rho)^{1/2}\Big/p\ c^2,$$
$$\eta^* = \eta(4\pi G\rho)^{1/2}\Big/c^2,\ \upsilon_0^* = \upsilon_0(4\pi G\rho)^{1/2}\Big/c^2,\ V^{*2} = V^2\Big/c^2\ ,\ \Omega^* = \Omega\Big/(4\pi G\rho)^{1/2}. \qquad (34)$$

Numerical calculations were performed to determine the roots of ω^* as a function of wave number k^* for several values of different parameters involved taking $\gamma = 5/3$. Out of four modes, only one mode is unstable for which the calculations are presented in Figures 1-3, where the growth rate ω^* (positive real values of ω) has been plotted against the wave number k^* to show the dependence of the growth rate on the different physical parameters such as porosity, rotation and FLR corrections.

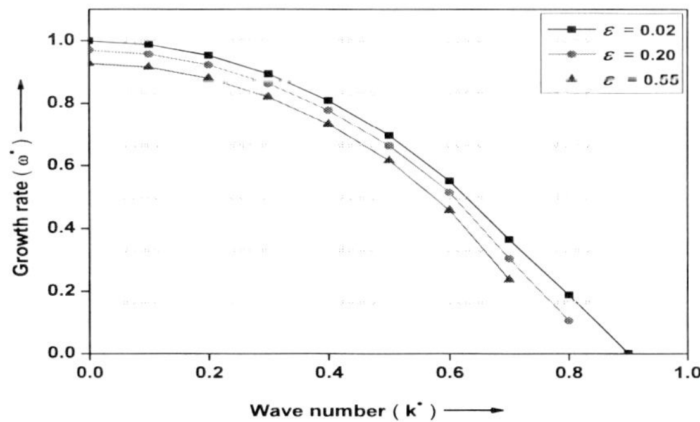

Figure. 1. Growth rate (positive values of w^*) against wave number k^* for three values of parameter $\varepsilon = 0.02,\ 0.2,\ 0.55,$ keeping the other parameters fixed $L_T^* = 1, L_\rho^* = 0, V^* = 1,\ \lambda^* = 1, v_0^* = 1,$ and $\Omega^* = 1.0$.

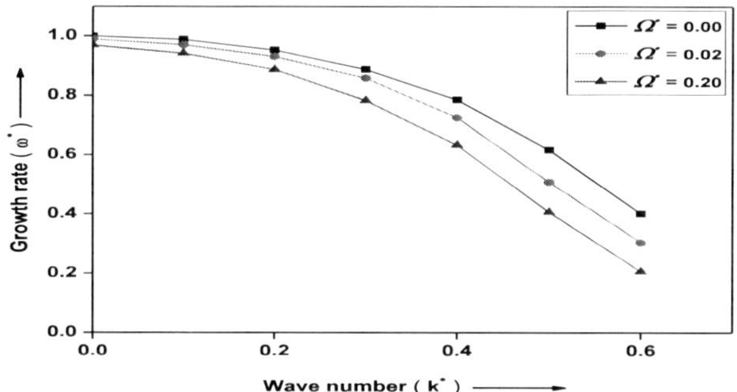

Figure. 2. Growth rate (positive values of w^*) against wave number k^* for three values of parameter $\Omega^* = 0.0,\ 0.02, 0.2,$ keeping the other parameters fixed $L_T^* = 1.0, L_\rho^* = 0,\ V^* = 1,\ \lambda^* = 1.0,\ v_0^* = 1,$ and $\varepsilon = 1.0.$

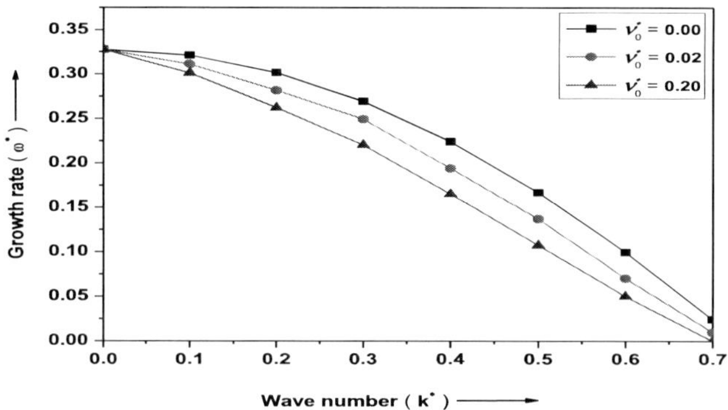

Figure. 3. Growth rate (positive values of w^*) against wave number k^* for three values of parameter $v_0^* = 0.0,\ 0.02, 0.2,$ keeping the other parameters fixed $L_T^* = 1.0, L_\rho^* = 0,\ V^* = 1, \lambda^* = 1.0, \Omega^* = 1,$ and $\varepsilon = 1.0.$

It is clear from Figure 1 that the peak value of the growth rate decreases with increase in the value of medium porosity. Thus the effect of medium porosity is stabilizing on the growth rate of the system. From Figure 2 we see that the growth rate decreases with increase in the value of rotation. Thus we conclude that rotation stabilize the growth rate of the system. One can observe from Figure 3 that the growth rate decreases with increasing FLR corrections. Thus the effect of FLR corrections is stabilizing on the growth rate of the system.

4.1.2. Axis of Rotation Perpendicular to the Magnetic Field (Ω ^ B)

In the case of axis of rotation perpendicular to the magnetic field, we put $\Omega_x = \Omega$ and $\Omega_z = 0$ in the dispersion relation (26) which reduces to

$$\omega\left\{\omega^2\left(\omega^2+4\varepsilon^2\Omega^2+\varepsilon^2\upsilon_0^2k^4\right)+\left(\omega^2+4\varepsilon^2\Omega^2\right)\left[\omega\frac{V^2k^2}{\omega+\eta k^2}+\frac{\Omega_I^2+\omega\Omega_J^2}{\omega+\beta}\right]\right\}=0. \qquad (35)$$

This dispersion relation represents the combined influence of FLR corrections, rotation, porosity magnetic field, radiative heat-loss function and thermal conductivity on the Jeans-gravitational instability of the considered system. In absence of FLR corrections this dispersion relation (35) is identical to that obtained by Prajapati et al. [41] neglecting viscosity and electron inertia in that case. On comparing our dispersion relation (35) with dispersion relation (34) of Prajapati et al. [41] we find that presence of FLR corrections coupled the rotation parameter with thermal and radiative mode. This is the new finding in our case than that of Prajapati et al. [41]. Equation (35) has two independent factors. The first factor of equation (35) gives $\omega = 0$, which is a marginal stable mode. The second factor of equation (35) gives the following dispersion relation on substituting the values of $\Omega_I^2, \Omega_J^2, \alpha$ and β.

In absence of rotation $(h = \Omega = 0)$ equation (35) becomes

$$\omega^3+\left[(\gamma-1)\left(\frac{T\rho L_T}{p}+\frac{\lambda k^2T}{p}\right)\right]\omega^2+\left[\varepsilon^2\upsilon_0^2k^4+V^2k^2+c^2k^2-4\pi G\rho\right]\omega+\left\{k^2(\gamma-1)\right.$$

$$\left.\times\left(TL_T-\rho L_\rho+\frac{\lambda k^2T}{\rho}\right)-(\gamma-1)\left(\frac{T\rho L_T}{p}+\frac{\lambda k^2T}{p}\right)\left[4\pi G\rho-\varepsilon^2\upsilon_0^2k^4-V^2k^2\right]\right\}=0. \qquad (36)$$

The condition of instability obtained from constant term of equation (36) is given as

$$\left\{k^2(\gamma-1)\left(TL_T-\rho L_\rho+\frac{\lambda k^2T}{\rho}\right)-(\gamma-1)\left(\frac{T\rho L_T}{p}+\frac{\lambda k^2T}{p}\right)\left[4\pi G\rho-\varepsilon^2\upsilon_0^2k^4-V^2k^2\right]\right\}<0. \quad (37)$$

On comparing equations (35) and (37) we find that condition of instability gets modified by FLR porosity corrections and porosity when medium is non-rotating. In present case condition of instability and growth rate of instability both depend on FLR corrections and porosity. In absence of FLR corrections and porosity the above inequality is identical to Aggarwal and Talwar [36] and also to Prajapati et al. [41] for ($f=1$). So we improve the result of Aggarwal and Talwar [36] and Prajapati et al. [41] for ($f=1$), by inclusion of FLR corrections and porosity in our present problem.

4.2. Longitudinal Mode of Propagation (K||B)

In this case the perturbations are taken to be parallel to the direction of the magnetic field ($i.e.,\ k_x=0,\ k_z=k$). The dispersion relation (25) reduces to

$$\left(\omega^2+\Omega_T^2\right)\left\{\omega\left[\omega+\frac{V^2k^2}{\omega}\right]^2+4\varepsilon^2\Omega_x^2\left[\omega+\frac{V^2k^2}{\omega}\right]+\omega\left(2\varepsilon\upsilon_0k^2-2\varepsilon\Omega_z\right)^2\right\}$$

$$-\Omega_T^2\left[\omega+\frac{V^2k^2}{\omega}\right]4\varepsilon^2\Omega_x^2=0. \quad (38)$$

We see that in the longitudinal mode of propagation the dispersion relation is modified due to the presence of FLR corrections, rotation, radiative heat-loss function, porosity and thermal conductivity. This dispersion relation (38) is further reduced for rotation axis parallel and perpendicular to the direction of the magnetic field.

4.2.1. Axis of Rotation along the Magnetic Field ($\Omega||B$)

For axis of rotation along the magnetic field, $\Omega_x = 0$ and $\Omega_z = \Omega$ equation (38) reduces to

$$\omega\left(\omega^2 + \frac{\Omega_I^2 + \omega\Omega_J^2}{\omega + B}\right) \times \left\{\left[\omega + \frac{V^2k^2}{\omega}\right]^2 + \left(2\varepsilon\upsilon_0 k^2 - 2\varepsilon\Omega\right)^2\right\} = 0. \quad (39)$$

This dispersion relation shows the combined influence of rotation, FLR corrections, thermal conductivity, radiative heat-loss function, porosity on the Jeans-gravitational instability of the hydromagnetic fluid plasma. In absence of FLR corrections and porosity this dispersion relation is identical to that of Prajapati et al. [41] neglecting Hall current, electron inertia and viscosity in that case. Equation (39) has three independent factors. The first factor of equation (39) gives $\omega = 0$, which is a marginal stable mode. The second factor of equation (39) gives the following dispersion relation on substituting the values of $\Omega_I^2, \Omega_J^2, \alpha$ and β.

$$\omega^3 + \left[(\gamma-1)\left(\frac{T\rho L_T}{p} + \frac{\lambda k^2 T}{p}\right)\right]\omega^2 + \left(c^2k^2 - 4\pi G\rho\right)\omega + \left[k^2(\gamma-1)\left(TL_T - \rho L_\rho + \frac{\lambda k^2 T}{\rho}\right) - 4\pi G\rho(\gamma-1)\left(\frac{T\rho L_T}{p} + \frac{\lambda k^2 T}{p}\right)\right] = 0. \quad (40)$$

The condition of instability obtained from constant term of equation (40) is given as

$$\left\{k^2(\gamma-1)\left(TL_T - \rho L_\rho + \frac{\lambda k^2 T}{\rho}\right) - 4\pi G\rho(\gamma-1)\left(\frac{T\rho L_T}{p} + \frac{\lambda k^2 T}{p}\right)\right\} < 0, \quad (41)$$

The above condition of radiative instability is independent of rotation, FLR corrections and porosity. This condition of radiative instability is also identical to that of Bora and Talwar [37]. But in the present problem we have considered the effects of rotation, porosity and FLR corrections, Bora and Talwar [37] have not considered these effects. So the dispersion relation and the growth rate of radiative instability in the present problem is modified due to the simultaneous inclusion of rotation, porosity and FLR corrections, but the condition of radiative instability is unaffected by the presence of rotation, porosity and FLR corrections. Thus we conclude that rotation, porosity and FLR corrections parameters does not influence the condition of radiative instability (41), but presence of rotation, porosity and FLR corrections modify the growth rate of instability of the condition (41). This is the new result, which was not earlier obtained by Bora and Talwar [37].

The third factor of equation (39) on simplification gives the following dispersion relation medium ($h = 0$) equation (39) becomes

$$\omega^4 + \left(2V^2k^2 + 4\varepsilon^2\Omega^2 + 4\varepsilon^2\upsilon_0^2k^4 - 8\varepsilon^2\Omega\upsilon_0k^2\right)\omega^2 + V^4k^4 = 0. \quad (42)$$

The above equation represents Alfven mode modified by the presence of rotation, porosity and FLR corrections. Equation (42) is modified form of equation (31) of Kaothekar and Chhajlani [20] by inclusion of rotation in the present problem. The roots of equation (42) are

$$\omega_{1,2}^2 = -\left(V^2k^2 + 2\varepsilon^2\upsilon_0^2k^4 + 2\varepsilon^2\Omega^2 - 4\varepsilon^2\Omega\upsilon_0k^2\right) \pm \left[\left(V^2k^2 2\varepsilon^2\upsilon_0^2k^4 + 2\varepsilon^2\Omega^2 - 4\varepsilon^2\Omega\upsilon_0k^2\right)^2 - V^4k^4\right]^{1/2}. \quad (43)$$

Hence rotation, porosity and FLR corrections modify the Alfven mode by changing the growth rate of the system.

4.2.2. Axis of Rotation Perpendicular to the Magnetic Field (Ω ^ B)

For axis of rotation perpendicular to the magnetic field, we substitute $\Omega_x = \Omega$ and $\Omega_z = 0$ in the dispersion relation (38) which reduces to give

$$\omega \left\{ 4\varepsilon^2\Omega^2\left[\omega^2 + \omega\frac{V^2k^2}{\omega}\right] + \left(\omega^2 + \frac{\Omega_I^2 + \omega\,\Omega_J^2}{\omega+\beta}\right) \times \left[4\varepsilon^2\upsilon_0^2k^4 + \left(\omega + \frac{V^2k^2}{\omega}\right)^2\right]\right\} = 0. \tag{44}$$

The dispersion relation shows the combined influence of rotation, porosity FLR corrections, radiative heat-loss function, and thermal conductivity on the Jeans-gravitational instability of the hydro-magnetic fluid plasma. In absence of FLR corrections and porosity this dispersion relation is identical to that of Prajapati et al. [41] neglecting electron inertia, Hall current and viscosity in their case. This dispersion relation is the product of two independent factors. The first factor $\omega = 0$, is a marginal stable mode. The second factor of equation (44) gives, on substituting the values of Ω_I^2, Ω_J^2, α and β the following seven degree polynomial equation:

$$\omega^7 + \left[(\gamma-1)\left(\frac{T\rho L_T}{p} + \frac{\lambda k^2 T}{p}\right)\right]\omega^6 + \left[4\varepsilon^2\Omega^2 + 4\varepsilon^2\upsilon_0^2k^4 + 2V^2k^2 + c^2k^2 - 4\pi G\rho\right]\omega^5$$

$$+(\gamma-1)\left(\frac{T\rho L_T}{p} + \frac{\lambda k^2T}{p}\right)\left(4\varepsilon^2\Omega^2 + 4\varepsilon^2\upsilon_0^2k^4 + 2V^2k^2\right) + k^2(\gamma-1)\left(TL_T - \rho L_\rho + \frac{\lambda k^2T}{\rho}\right)$$

$$-4\pi G\rho(\gamma-1)\left(\frac{T\rho L_T}{p} + \frac{\lambda k^2T}{p}\right)\Bigg]\omega^4 + \Big[(c^2k^2 - 4\pi G\rho) + 4\varepsilon^2\upsilon_0^2k^4\left(c^2k^2 - 4\pi G\rho\right) + 2V^2k^2$$

$$\times\left(2\varepsilon^2\Omega^2 + c^2k^2 - 4\pi G\rho\right) + V^4k^4 - 4\pi G\rho\left(\frac{T\rho L_T}{p} + \frac{\lambda k^2T}{p}\right)\Bigg]\Bigg]\omega^3 + +(\gamma-1)\left(\frac{T\rho L_T}{p} + \frac{\lambda k^2T}{p}\right)$$

$$\times V^2k^2\left(4\varepsilon^2\Omega^2 + V^2k^2\right) + \left(4\varepsilon^2\upsilon_0^2k^4 + 2V^2k^2\right)\left[k^2(\gamma-1)\left(TL_T - \rho L_\rho + \frac{\lambda k^2T}{\rho}\right) - 4\pi G\rho\right.$$

$$\times(\gamma-1)\left(\frac{T\rho L_T}{p}+\frac{\lambda k^2 T}{p}\right)\Bigg]\Bigg]\omega^2+\Big[\big(c^2k^2-4\pi G\rho\big)\big(V^4k^4\big)+\big(V^4k^4\big)\Big[k^2(\gamma-1)\left(TL_T-\rho L_\rho+\frac{\lambda k^2 T}{\rho}\right)$$

$$-4\pi G\rho(\gamma-1)\left(\frac{T\rho L_T}{p}+\frac{\lambda k^2 T}{p}\right)\Big]\Big]=0. \quad (45)$$

The above dispersion relation represents the effect of porosity radiative heat-loss function, thermal conductivity, rotation and FLR corrections on the Jeans-gravitational instability of magnetized homogeneous plasma for longitudinal mode of propagation with axis of rotation perpendicular to magnetic field. The condition of instability obtained from constant term of equation (45) is given as

$$\left\{k^2(\gamma-1)\left(TL_T-\rho L_\rho+\frac{\lambda k^2 T}{\rho}\right)-4\pi G\rho(\gamma-1)\left(\frac{T\rho L_T}{p}+\frac{\lambda k^2 T}{p}\right)\right\}<0. \quad (46)$$

The above condition of instability is identical to the condition (47) and discussed there. We find that for both the cases of rotation parallel and perpendicular to the direction of magnetic field, the condition of instability obtained is same. We also find that the condition of instability for the case of longitudinal mode of propagation in both the cases of rotation is the same and there is no effect of rotation and FLR corrections on this condition of instability. Thus the above condition of instability is independent of the direction of rotation and depends only on radiative heat-loss function and thermal conductivity of the medium. To discuss the stability of the system, if the condition (46) is not satisfied, then according to Routh-Hurwitz criterion the system will necessarily be stable. To satisfy the necessary condition all the coefficients of equation (45) must be positive, and for the sufficiency of the condition the principal diagonal minors of Hurwitz matrix must also be positive on calculating all the minors, we find that all the Hurwitz minors are positive satisfying the sufficient condition for stability of the system.

To see the effect of rotation, porosity and FLR corrections, we analyze the dispersion relation (45) and find that coefficients of this dispersion relation are having terms associated with rotation and FLR corrections and if we plot the growth rate of this dispersion relation by solving it numerically, then the growth rate of the unstable mode will be modified by rotation, porosity and FLR corrections. So we conclude that the condition of instability given by equation (46) is independent of FLR corrections, porosity and rotation, but the growth rate of the system depends on rotation and FLR corrections.

In astrophysical situation, instability of the system is one of the most important causes of configuration of objects. So we study the effects of medium porosity ε, rotation Ω^*, and FLR corrections υ_0^* on the growth rate of unstable mode. We present the dispersion relation (45) in a non-dimensional form on dividing it by $(4\pi G\rho)^2$ and the value of γ in numerical calculations is taken as 5/3.

$$\omega^{*7}+\left[\left(L_T^*+\lambda^* k^{*2}\right)\right]\omega^{*6}+\left[4\varepsilon^2\Omega^{*2}+4\varepsilon^2\upsilon_0^{*2}k^{*4}+2V^{*2}k^{*2}+k^{*2}-1\right]\omega^{*5}$$
$$+\left(L_T^*+\lambda^* k^{*2}\right)\left(4\varepsilon^2\Omega^{*2}+4\varepsilon^2\upsilon_0^{*2}k^{*4}+2V^{*2}k^{*2}\right)k^{*2}\left\{\frac{1}{\gamma}\left(L_T^*+\lambda^* k^{*2}\right)-L_\rho^*\right\}$$
$$-\left(L_T^*+\lambda^* k^{*2}\right)\Big]\omega^{*4}+4\varepsilon^2\upsilon_0^2k^{*4}\left(k^{*2}-1\right)+2V^{*2}k^{*2}\left(2\varepsilon^2\Omega^{*2}+k^{*2}-1\right)+V^{*4}k^{*4}$$
$$-\left(L_T^*+\lambda^* k^{*2}\right)\Big]\omega^{*3}+\Big[\left(L_T^*+\lambda^* k^{*2}\right)V^{*2}k^{*2}\left(4\varepsilon^2\Omega^{*2}+V^{*2}k^{*2}\right)+\left(L_T^*+\lambda^* k^{*2}\right)V^{*2}k^{*2}$$
$$\times\left(4\varepsilon^2\Omega^{*2}+V^{*2}k^{*2}\right)+\left(4\varepsilon^2\upsilon_0^{*2}k^{*4}+2V^{*2}k^{*2}\right)+k^{*2}\left\{\frac{1}{\gamma}\left(L_T^*+\lambda^* k^{*2}\right)-L_\rho^*\right\}$$
$$-\left(L_T^*+\lambda^* k^{*2}\right)\Big]\omega^{*2}+\left[\left(k^{*2}-1\right)\left(V^{*4}k^{*4}\right)\right]\omega^*+\left(V^{*4}k^{*4}\right)$$
$$\times\left[k^{*2}\left\{\frac{1}{\gamma}\left(L_T^*+\lambda^* k^{*2}\right)-L_\rho^*\right\}-\left(L_T^*+\lambda^* k^{*2}\right)\right]=0. \tag{47}$$

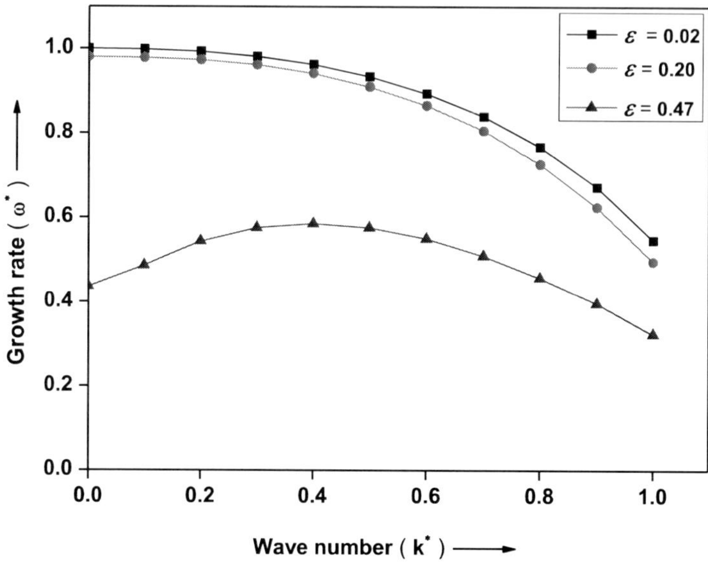

Figure 4. Growth rate (positive values of w^*) against wave number k^* for three values of parameter $\varepsilon = 0.02,\ 0.2, 0.47,$ keeping the other parameters fixed $L_T^* = 1.0, L_\rho^* = 0,\ V^* = 1,\ \lambda^* = 1.0, \nu_0^* = 1,$ and $\Omega^* = 1.0$.

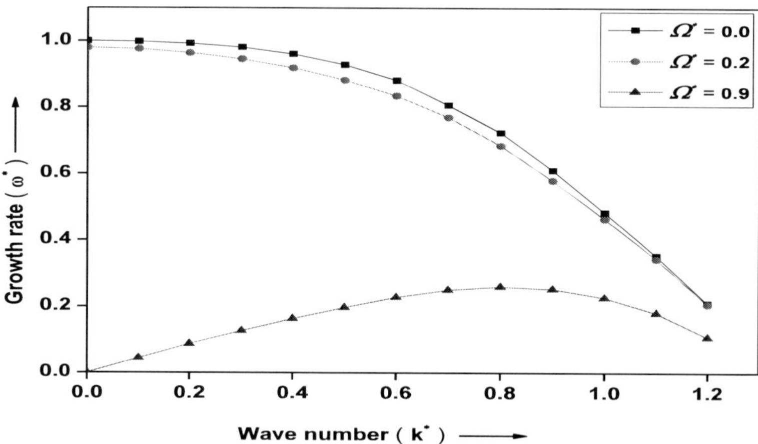

Figure 5. Growth rate (positive values of w^*) against wave number k^* for three values of parameter $\Omega^* = 0.0,\ 0.2, 0.9,$ keeping the other parameters fixed $L_T^* = 1.0,\ L_\rho^* = 0,\ V^* = 1,\ \lambda^* = 1.0,\ \nu_0^* = 1,$ and $\varepsilon = 1.0$.

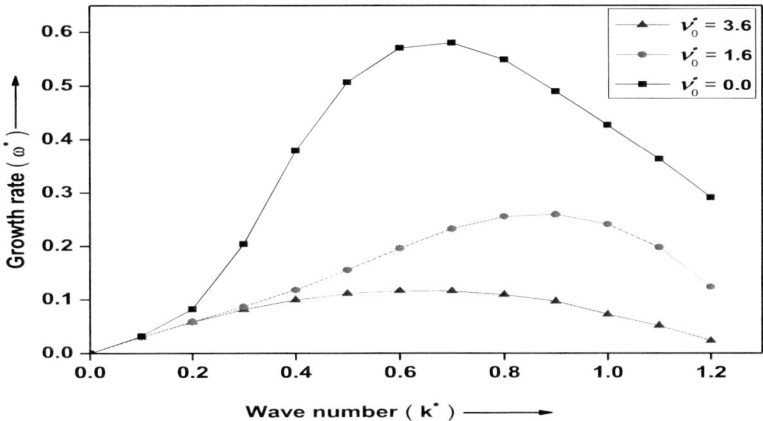

Figure 6. Growth rate (positive values of w^*) against wave number k^* for three values of parameter $v_0^* = 0,\ 0.17, 3.7,$ keeping the other parameters fixed $L_T^* = 1.0, L_\rho^* = 0,\ V^* = 1, \lambda^* = 1.0, \Omega^* = 1,$ and $\varepsilon = 1.0.$

Numerical calculations were performed to determine the roots of ω^* as a function of wave number k^* for several values of different parameters involved taking $\gamma = 5/3$. Out of four modes, only one mode is unstable for which the calculations are presented in Figures 1-3, where the growth rate ω^* (positive real values of ω) has been plotted against the wave number k^* to show the dependence of the growth rate on the different physical parameters such as porosity, rotation and FLR corrections.

It is clear from Figure 4 that the peak value of the growth rate decreases with increase in the value of medium porosity. Thus the effect of medium porosity is stabilizing on the growth rate of the system. From Figure 5 we see that the growth rate decreases with increase in the value of rotation. Thus we conclude that rotation stabilize the growth rate of the system. One can observe from Figure 6 that the growth rate decreases with increasing FLR corrections. Thus the effect of FLR corrections is stabilizing on the growth rate of the system.

CONCLUSION

In the present problem we have carried out the effect of rotation, porosity and FLR corrections on the Jeans-gravitation instability of plasma including the effects of radiative heat-loss function and thermal conductivity. The general dispersion relation is obtained, which is modified due to the presence of considered physical parameters. This dispersion relation is reduced for transverse and longitudinal wave propagation to the direction of magnetic field, which is further discussed for rotation axis parallel and perpendicular to the direction of magnetic field.

In the case of transverse wave propagation to the direction of magnetic field with axis of rotation along magnetic field we obtained two modes. The first mode is a marginal stable mode. The second mode gives the gravitating thermal mode influenced by rotation, porosity FLR corrections and radiative heat-loss function. We find that the condition of gravitational instability is modified due to the presence of porosity rotation, FLR corrections, radiative heat-loss function, and thermal conductivity. For the case of non-FLR medium, it is observed that the condition of radiative instability and expression of critical Jeans wave number both are modified due to the presence of porosity, rotation, and it shows the stabilizing influence. It is observed that for non-FLR the condition of radiative instability and expression of critical Jeans wave number both are modified due to the presence of porosity, rotation and magnetic field. It is independent of FLR corrections.

In the case of axis of rotation perpendicular to the magnetic field for transverse wave propagation, we obtained two modes. The first mode is a marginal stable mode. The second mode represents the effect of porosity rotation, FLR corrections, radiative heat-loss function and thermal conductivity on Jeans-gravitational instability of plasma. We find that the condition of instability is independent of porosity, rotation and FLR corrections and it depends on radiative heat-loss function and thermal conductivity. But the growth rate of the system is affected by the presence of rotation, porosity and FLR corrections. For the case of non-rotating medium, it is observed that condition of radiative instability is modified by the presence of FLR corrections, porosity and magnetic field, and it shows stabilizing influence.

ACKNOWLEDGMENTS

Author (S.K.) is grateful to Er. Praveen Vashishtha, Chairman Mahakal Institute of Technology & Management Ujjain and Dr. J. N. Vyas Director, Mahakal Institute of Technology & Management Ujjain for continuous support.

REFERENCES

[1] Chandrasekhar, S. *Hydrodynamics and Hydromagnetic Stability*, Oxford:

[2] Houser, J. L. (1998). The effect of rotation on the gravitational radiation and dynamical stability of stiff stellar core, *Mon. Not. R. Astron. Soc.*, vol. 299, 1069.

[3] Soni, G. D. & Chhajlani, R. K. (1998). Magnetogravitational instability of a rotating anisotropic plasma with FLR corrections and generalized polytropic laws, *J. Plasma Phys.*, vol. 60, 673.

[4] Tandon, J. N. & Talwar, S. P. (1963). Stability of rotating plasma with anisotropic pressure, *Nucl. Fusion*, vol. 3, 75.

[5] Prajapati, R. P., Bhakta, S. & Chhajlani, R. K. (2008). Jeans instability in coalitional strongly coupled dusty plasma with radiative condensation and polarization force, *Phys. Plasmas*, vol. 23, 053703.

[6] Uberoi, C. (2009). Electron inertia effects on the transverse gravitational instability, *J. Plasma Fusion Res. Series*, vol. 8, 823.

[7] Vyas, M. K. & Chhajlani, R. K. (1988). Magnetogravitational instability of an infinite homogeneous rotating plasma through a porous medium with finite electrical and thermal conductivities, *Astrophys. Space Sci.,* vol. 140, 89.

[8] Khan, A. & Bhatia, P. K. (1993). Gravitational instability of a rotating fluid in an oblique magnetic field, *Phys. Scr.*, vol. 47, 230.

[9] Prajapati, R. P., Soni, G. D. & Chhajlani, R. K. (2008). Self-gravitational rotating anisotropic heat conducting plasma, *Phys. Plasmas*, vol. 15, 012107.

[10] Nield, D. A. & Bejan, A. (1999). *Convection in Porous Media*, 2nd ed., Berlin: Springer.

[11] Vafai, K., (2000). *Handbook of Porous Media*, New York: Marcel Dekker, Clarendon Press.

[12] Vyas, M. K. & Chhajlani, R. K. (1988). Effect of thermal conductivity on the gravitational instability of a magnetized rotating plasma through a porous medium in the presence of suspended particles, *Astrophys. Space Sci.*, vol. 145, 223.

[13] Sharma, R. C. & Thakur, K. P. (1982). The instability of flow through porous medium for some systems of astrophysical interest, *Astrophys. Space Sci.*, 81, 95.

[14] Kaothekar, S. & Chhajlani, R. K. (2012). Effect of radiative heat-loss function and finite Larmor radius corrections on Jeans instability of viscous thermally conducting self-gravitating astrophysical plasma, *Astron. Astrophys.,* 2012, 420938.

[15] El-Sayed, M. F. & Mohamed, R. A. (2010). Magnetogravitational instability of thermally conducting rotating viscoelastic fluid with Hall current in Brinkman porous medium, *J. Porous Media*, vol. 13, 779.

[16] Rana. G. C. (2013). Thermoslutal convectionin Walter's (modle B') rotating fluid permeated with suspended particles and variable

gravity field in porous medium in hydromagnetics, *J. Applied Fluid Mechanics*, 6, 87.

[17] El-Sayed, M. F. & Hussein, D. F. (2013). Gravitational instability of thermally conducting fluid in a variable magnetic field through porous medium with Hall current, *J. Natural Sci. Math.*, 6, 89.

[18] Chand, R. & Rana, G. C. (2014). The effects of radiation on the onset of thermal instability in a layer of nanofluid layer in a porous medium, *Int. J, Appl. Math, and Mech.*, 10, 76.

[19] Chand, R., Rana, G. C. & Singh, K. (2015). Thermal instability in a Rivlin-Ericksen elastico-viscous nanofluid in aporous medium: a revised model, *Int. J. Nanosci. Nanoengg.*, 2, 1.

[20] Kaothekar, S. & Chhajlani, R. K. (2013). Jeans instability of self gravitating partially ionized Hall plasma with radiative heat loss functions and porosity, *J. Phys. Con. Series.*, 1536, 1288.

[21] Jukes, J. D. (1964). Gravitational resistive instability in plasma with finite Larmor radius, *Phys. Fluids*, 7, 52.

[22] Roberts, K. V. & Taylor, J. B. (1962). Magnetohydrodynamic equations for finite Larmor radius, *Phys. Rev. Lett.*, 8, 197.

[23] Rosenbluth, M. N., Krall, N. & Rostoker, N. (1962). Finite Larmor radius stabilization of "weakly" unstable confined plasmas, *Nucl. Fusion Suppl.*, 1, 143.

[24] Singh, S. & Hans, H. K. (965). Effect of ion gyration radius on magnetogravitatiolal instability of an infinite homogeneous medium, *Zeit. Astrophys.*, 62, 12.

[25] Herrnegger, F. (1972). Effect of collisions and gyroviscosity on gravitational instability in a two-component plasma, *J. Plasma Phys.*, 8, 393.

[26] Sharma, R. C. (1974). Gravitational instability of a rotating plasma, *Astrophys. Space Sci.*, 29, L1.

[27] Chhonkar, R. P. S. & Bhatia, P. K. (1977). Larmor radius effects on the gravitational instability of a two-component plasma, *J. Plasma Phys*, 18, 273.

[28] Devlen, E. & Pekunlu, E. R. (2010). Finite Larmor radius effects on weakly magnetized, dilute plasmas, *Mon. Not. R. Astron. Soc.*, 404, 830.

[29] Kaothekar, S. & Chhajlani, R. K. (2013). Effect of porosity and FLR corrections on Jeans instability of self-gravitating radiative thermally conducting viscous plasma, *J. Porous Media*, 16, 709.

[30] Field, G. B. (1965). Thermal instability, *Astrophys. J.*, *142*, 531.

[31] Hunter, J. H. (1966). The role of thermal instabilities in star formation, *Mon. Not. R. Astron. Soc.*, 133, 239.

[32] Ibanez, M. H. S. (1985). Sound and thermal waves in a fluid with an arbitrary heat-loss function, *Astrophys. J.*, 290, 33.

[33] Talwar, S. P. & Bora, M. P. (1995). Thermal instability in a star-gas system, *J. Plasma Phys.*, 54, 157.

[34] Burkert, A. & Lin, D. N. C. (2000). Thermal instability and the formation of clumpy gas clouds, *Astrophys. J.*, 537, 270.

[35] Radwan, A. E. (2004). Variable streams selg-gravitating instability of radiating rotating gas cloud, *Applied Mathematics and Computation*, 148, 331.

[36] Aggarwal, M. & Talwar, S. P. (1969). Magneto-thermal instability in a rotating gravitating fluid, *Mon. Not. R. Astro. Soc.*, 146, 235.

[37] Bora, M. P. & Talwar, S. P. (1993). Magnetothermal instability with generalized Ohm's law, *Phys. Fluids B*, 5, 950.

[38] Prajapati, R. P., Parihar, A. K. & Chhajlani, R. K. (2008). Self-gravitational rotating anisotropy pressure plasma in presence of Hall current and electrical resistivity using generalized polytrope laws, *Phys. Plasmas*, 15, 062108.

[39] Kaothekar, S., Soni, G. D. & Chhajlani, R. K. (2012). Effect of neutral collision and radiative heat-loss function on self-gavitationa instability of viscous thermally conducting partially-ionized plasma, *AIP Adv.*, 2, 042191.

[40] Sharma, P. & Jain, S. (2016). Radiative condensation instability in partially ionized dusty plasma with polarization force, *Phys. Scr*, 23, 015602.

[41] Prajapati, R. P., Pensia, R. K., Kaothekar, S. & Chhajlani, R. K. (2010). Self-gravitational instability of rotating viscous Hall plasma with arbitrary radiative heat-loss functions and electron inertia, *Astrophys. Space Sci.*, 327, 139.

[42] Kaothekar, S. (2018). Thermal instability of partially ionized viscous plasma with Hall effect FLR corrections flowing through porous medium, *J. Porous Media*, 21, 679.

[43] Kaothekar, S. (2020). Jeans instability of rotating plasma with radiative heat-loss function and FLR corrections flowing through porous medium, *Astrophys. Space Sci.*, vol. 365, 1.

In: An Introduction to Molecular Clouds
Editor: Sachin Kaothekar
ISBN: 978-1-53619-178-3

Chapter 5

Molecular Cloud Formation via Thermal Instability of Viscous Partially-Ionized Plasma with Neutral Collision and Radiative Heat-Loss Function in an Interstellar Medium

G. D. Soni[1,*] and Sachin Kaothekar[2]

[1]Government Science College, Dewas, Madhya Pradesh, India
[2]Department of Physics, Mahakal Institute of Technology & Management, Ujjain, Madhya Pradesh, India

Abstract

The problem of thermal instability is investigated for a partially ionized plasma which has connection in astrophysical condensations for

[*] Corresponding Author's Email: gdsoniphysics@gmail.com.

the formation of molecular clouds in interstellar medium (ISM). The normal mode analysis technique is used in this problem. The general dispersion relation is obtained from linearized perturbation equations of the problem. Effects of collisions with neutrals, radiative heat-loss function, viscosity, thermal conductivity and magnetic field strength, on the thermal instability of the system are discussed. The conditions of instability are obtained for heat-loss function with thermal conductivity. Numerical computations have been performed to calculate the effect of different physical parameters on the growth rate of the thermal instability of considered system. The heat-loss function, thermal conductivity, viscosity, magnetic field and neutral collision have stabilizing effect, while finite electrical resistivity has a destabilizing effect on the growth rate of the thermal instability. Routh-Hurwitz's criterion is used to calculate, the stability of the system. The results presented in this problem are helpful in understanding the molecular cloud formation and star formation in ISM.

Keywords: molecular cloud formation, ISM (inter stellar medium), thermal instability, neutral collision frequency, radiative heat-loss functions

1. INTRODUCTION

One of the universally established facts is that the thermal and radiative processes play a significant role in instability investigations in thermally unstable astrophysical plasma. Thermal instability occurs when a positive temperature perturbation is completed in a thermal unstable medium; the perturbation produces and the growth rate diminish. This development is thought to be possible in a number of astrophysical situations such as the gas in clusters of the galaxies, in the solar corona and in the interstellar medium. Thermal instability has many applications in astrophysical situations (e.g., a clumpy interstellar medium, stellar atmosphere, star formation, globular clusters and galaxy formation and many more situations). In this connection a number of authors examined the problem of thermal instability occur due to heat-loss process in plasma. The thermal instability arising due to various heat-loss processes in a dilute plasma may be a possible cause of astrophysical condensations and the formation of large and small objects (Field [1], Hunter [2], Cook et al. [3]). Apart from these, there are verities of astrophysical circumstances where the effect of thermal instability is important. Van Hoven & Mok [4] have examined the consequence of thermal instability in a sheared magnetic field. The different works have argued the effect of thermal instability in the fragmentation of thermal fluids (Hunter [5], Aggarwal & Talwar [6], [7]). Thermal instabilities in case of active galactic nuclei have been calculated by Beltrametti [8]. Gupta et al. [9] have acarried out the problem of thermal instability in a high temperature rotating and gravitating plasma. Ibanez [10] has calculated the sound and thermal waves in a fluid with an arbitrary heat-loss function. Kim & Narayan [11] have argued the thermal instability in clusters of galaxies with conduction incorporating the consequence of radiative heat loss function. Radwan [12] has calculated the self-gravitating instability of radiating rotating gas cloud streams with non-uniform velocity. Menou et al. [13] gave the significance of radiative effects in the Sun's upper radiative zone. Inutsuka et al. [14] have calculated the transmit of shock wave into warm neutral medium taking into account radiative heating and cooling, thermal conduction and viscosity. Shadmehri & Dib [15] have investigated the thermal instability in magnetized partially ionized plasma with charged dust particles and radiative cooling function. Bora & Talwar [16] have examined the consequence of thermal and radiative instability in magnetized plasma with Hall current and electron inertia. Renard & Chieze [17] have carried out the thermal analysis of thermal stability to a molecular gas subjected to thermal exchanges of the interstellar medium. Dwivedi et al. [18] have examined the effect of radiative condensation on thermal instability. Talwar & Bora [19] have carried out the thermal instability in a star-gas system. Prajapati et al. [20] have explored the effect of radiative heat-loss function and thermal conductivity on thermal instability of fully ionized plasma with electron inertia, Hall current, rotation and viscosity. Hobbs et al. [21] have examined thermal instability in cooling galactic coronae: fuelling star formation in galactic disc. Dutta [22] has investigated the problem of the chemothermal instability have any role in the fragmentation of primordial gas. Linga & Tom [23] have discussed star forming filaments in warm dark metal models. Khesali et al. [24] have examined thermal instability in molecular cloud including dust particles, Hall effect and ambipolar diffusion. Voit [25] have carried out the problem of regulation of star formation in giant galaxies by precipitation, feedback and conduction. Wareing et al. [26] have examined mangnto-hydrodynamic simulations of stellar feedback in a sheet-like molecular cloud formed by thermal instability. McNamara [27] has investigated a mechanism for stimulated ANG feedback in massive galaxies. The analysis of problem of thermal instability and thermal conductivity has been carried out by some investigators (Ali & Bhatia [28], Bhatia & Hazarika [29] and Shaikh et al. [30]). Jain et al. [31] have explored the problem of effect of finite Larmor radius corrections on the thermal instability of thermally conducting viscous plasma with Hall current and electron inertia. Recently Kaothekar [32] has explored the problem of molecular cloud formation via thermal instability of finite resistive viscous radiating plasma with finite Larmor radius corrections. More recently Kaothekar [33] has investirated the problem of thermal instability of partially ionized viscous plasma with Hall effect FLR corrections flowing through porous medium. All these researchers have argued the problem with different parameters, but no one argued the collective effect of all the parameters together with radiative consequences.

In all the above mentioned study's authors have measured the intermediate as fully ionized. It is well recognized that a diminutive portion of interstellar gas may survive in the appearance of ionized tenuous, luminous and ionized gas identified as HI and HII region and the take it easy is largely neutral hydrogen. Thus the interstellar gas is not fully ionized but is saturated with neutral atoms. HI district and molecular clouds are thin ionized medium. Imperceptibly and fully ionized plasmas exist side by side in an assortment of sections in the universe. A real physical situation of interstellar medium put forwards that the interstellar plasma is a assortment of ionized constituent and a neutral component, and both the components interact through common collisions. In cosmic atmosphere these types of circumstances originate in chilly interstellar clouds, chromospheres and in the solar photosphere. The analysis of problem of thermal instability and thermal conductivity has been accepted out by a quantity of examiners (Ali & Bhatia [28], Bhatia & Hazarika [29] and Shaikh et al. [30]). All these investigators have conversed the difficulty with dissimilar limitations, but no one conversed the collective effect of all the constraints jointly with radiative effects. The results discussed in this problem are helpful for understanding of molecular cloud formation in interstellar medium (ISM).

Thus the plasma MHD model could be written including the effect of neutral particles, viscosity, thermal conductivity, finite electrical resistivity and radiative effects. Further, in the previous works, in general only the condition of instability is discussed to find the effect of various factors on gas condensation in thermal instability problem, but for description of the problem stability of the system should also be investigated. The object of the work presented in this paper is to discuss instability as well as stability of an infinite homogeneous, uniformly magnetized, thermally conducting, partially-ionized viscous plasma including finite electrical resistivity and radiative effects. Routh-Hurwitz's criterion is applied to discuss the stability of the system.

2. Basic Set of Equations of the Problem

Let us think the movement of an infinite homogeneous, thermally conducting, radiating, finitely conducting, viscous plasma having ionized component of density ρ and neutral particles of density ρ_d.

The uniform magnetic field B (0, 0, B) interacts only with the ionized component of the plasma. The density of the neutral particles is assumed to be small enough $(\rho_d << \rho)$ to neglect it contribution in pressure gradient of the neutrals. Also the viscosity of neutral is neglected, and we neglect the contribution of ion-neutral resistivity in induction equation. The equilibrium velocities of both the component are assumed equal to zero. In the momentum transfer equation of neutral the effect of pressure is neglected. Thus we are discussing the mutual frictional effects amid the neutral gas and the ionized fluid. The equations of the problem are written as

$$\rho \frac{d\boldsymbol{u}}{dt} = -\nabla p + \frac{1}{4\pi}(\nabla \times \boldsymbol{B}) \times \boldsymbol{B} + \rho_d v_c (\boldsymbol{u}_d - \boldsymbol{u}) + \rho v [\nabla^2 \boldsymbol{u} + \frac{1}{3}\nabla(\nabla.\boldsymbol{u})], \quad (1)$$

$$\frac{d\boldsymbol{u}_d}{dt} = -v_c (\boldsymbol{u}_d - \boldsymbol{u}), \quad (2)$$

$$\frac{d\rho}{dt} = -\rho \nabla.\boldsymbol{u}, \quad (3)$$

$$\frac{1}{\gamma - 1}\frac{dp}{dt} - \frac{\gamma}{\gamma - 1}\frac{p}{\rho}\frac{d\rho}{dt} + \rho\zeta - \nabla.(K\nabla T) = 0, \quad (4)$$

$$p = \rho R T, \quad (5)$$

$$\frac{\partial \boldsymbol{B}}{\partial t} = \nabla \times (\boldsymbol{u} \times \boldsymbol{B}) + \eta \nabla^2 \boldsymbol{B}, \quad (6)$$

$$\nabla.\boldsymbol{B} = 0, \quad (7)$$

where u, u_d, p, T, v, and K, represent respectively the velocity of ionized fluid, velocity of neutral gas, pressure, temperature, viscosity, thermal conductivity, and η, R, γ, ζ, and v_c are finite electrical resistivity, gas constant, ratio of two specific heats, heat-loss function, and collisional frequency between ionized and neutral components. The heat-loss function $\zeta(\rho,T)$ is the net loss of energy per unit mass of material per unit time, exclusive of thermal conduction, and in general it is a function of the local density and temperature of the gas. Operator d/dt is the substantial derivative given by

$$d/dt = (\partial/\partial t) + \boldsymbol{u}\,.\nabla\,. \tag{8}$$

In the equilibrium condition the fluid is assumed to be at rest. We assume a small amplitude perturbation with an oscillatory motion and as this perturbation grows in time the system is said to be unstable. The perturbations in density, pressure, magnetic field, ionized fluid velocity, temperature, and the heat-loss function are given as $\delta\rho$, δp, $\delta\boldsymbol{B}(\delta B_x$, δB_y, δB_z), $u(u_x, u_y, u_z)$, δT and $\partial\zeta$ respectively. The perturbation is given as

$\rho = \rho_0 + \delta\rho$, $\mathrm{p} = p_0 + \delta p$, $B = B_0 + \delta\boldsymbol{B}$, $u - u_0 + u$
(with $u_0 = 0$),
$T = T_0 + \delta T$, and $\zeta = \zeta_0 + \partial\zeta$
(with $\zeta_0 = 0,$), (9)

where the initial equilibrium is denoted by suffix '0'.

3. LINEARIZED PERTURBATION EQUATIONS

Substituting equation (10) in equations (1) to (7) and linearizing them, we get the linearized perturbation equations of the system neglecting higher order perturbations. We have dropped the suffix '0' from the equilibrium quantities. The linearized perturbation equations of the partially ionized plasma are

$$\rho\frac{\partial \boldsymbol{u}}{\partial t}=-\nabla\delta p+\frac{1}{4\pi}(\nabla\times\delta\boldsymbol{B})\times\boldsymbol{B}+\rho_d v_c(\boldsymbol{u}_{\mathrm{d}}-\boldsymbol{u})+\rho v[\nabla^2\boldsymbol{u}+\frac{1}{3}\nabla(\nabla.\boldsymbol{u})], \quad (10)$$

$$\frac{\partial \boldsymbol{u}_d}{\partial t}=-v_c(\boldsymbol{u}_{\mathrm{d}}-\boldsymbol{u}), \quad (11)$$

$$\frac{\partial\delta\rho}{\partial t}=-\rho\nabla.\boldsymbol{u}, \quad (12)$$

$$\frac{1}{\gamma-1}\frac{\partial\delta p}{\partial t}-\frac{\gamma}{\gamma-1}\frac{p}{\rho}\frac{\partial\delta\rho}{\partial t}+\rho[\zeta_\rho\delta\rho+\zeta_T\delta T]-K\nabla^2\delta T=0, \quad (13)$$

$$\frac{\delta p}{p}=\frac{\delta T}{T}+\frac{\delta\rho}{\rho}, \quad (14)$$

$$\frac{\partial\delta\boldsymbol{B}}{\partial t}=\nabla\times(\boldsymbol{u}\times\boldsymbol{B})+\eta\nabla^2\delta\boldsymbol{B}, \quad (15)$$

$$\nabla.\delta\boldsymbol{B}=0, \quad (16)$$

where ζ_ρ and ζ_T respectively denote partial derivatives $(\partial\zeta/\partial\rho)_T$ and $(\partial\zeta/\partial T)_\rho$ of the heat-loss function $\zeta(\rho,T)$, calculated for the equilibrium state.

4. Dispersion Relation

We seek solutions of equations (10) to (16) in which dependence of perturbed quantities is given by

$$\exp\,(ik\sin\theta\, x + ik\cos\theta\, z + i\sigma t)\,, \tag{17}$$

where σ denotes the frequency and k = ($k\sin\theta,\ 0,\ k\cos\theta$) in x and z directions respectively is the wave number of perturbation making angle θ with z-axis, such that

$$k^2 = k^2\sin^2\theta + k^2\cos^2\theta\,. \tag{18}$$

Equations (13) and (14) yield a relation between δp and $\delta\rho$ written as

$$\delta p = \left(\frac{A+\omega c^2}{\alpha+\omega}\right)\delta\rho\,, \tag{19}$$

where $\omega = i\sigma,\ \ c = (\frac{\gamma p}{\rho})^{1/2},\ \ A = (\gamma-1)(T\zeta_T - \rho\zeta_\rho + \frac{Kk^2T}{\rho}),$

$$\alpha = (\gamma-1)(\frac{T\rho}{p}\zeta_T + \frac{Kk^2T}{p})\,, \tag{20}$$

Substitution of equation (17) in equation (15) yields

$$iBk\cos\theta u_x - (\omega+\eta k^2)\delta B_x = 0\,, \tag{21}$$

$$iBk\cos\theta u_y - (\omega+\eta k^2)\delta B_y = 0\,, \tag{22}$$

$$iBk\sin\theta u_x + (\omega + \eta k^2)\delta B_z = 0\,. \tag{23}$$

Using equations (11) to (14) and (17) to (20) in equation (10), we find the following equations

$$\omega\{\omega(1 + \frac{\beta\nu_c}{\omega + \nu_c}) + D\}u_x + su_z - \frac{iBk\cos\theta}{4\pi\rho}\delta B_x + \frac{iBk\sin\theta}{4\pi\rho}\delta B_z = 0 \tag{24}$$

$$\omega\{\omega(1 + \frac{\beta\nu_c}{\omega + \nu_c}) + \nu k^2\}u_y - \frac{iBk\cos\theta}{4\pi\rho}\delta B_y = 0\,, \tag{25}$$

$$su_x + \{\omega(1 + \frac{\beta\nu_c}{\omega + \nu_c}) + J\}u_z = 0\,, \tag{26}$$

where,

$$\left.\begin{aligned} D &= \nu(k^2 + \frac{k^2\sin^2\theta}{3}) + \frac{k^2\sin^2\theta}{\omega}\{\frac{\Omega_I^2 + \omega\Omega_J^2}{k^2(\alpha + \omega)}\},\ \Omega_I^2 = k^2 A, \\ J &= \nu(k^2 + \frac{k^2\cos^2\theta}{3}) + \frac{k^2\cos^2\theta}{\omega}\{\frac{\Omega_I^2 + \omega\Omega_J^2}{k^2(\alpha + \omega)}\},\ \Omega_J^2 = c^2k^2 \end{aligned}\right\} \tag{27}$$

$$s = k^2\sin\theta\cos\theta\{\frac{\nu}{3} + \frac{1}{\omega}(\frac{\Omega_I^2 + \omega\Omega_J^2}{k^2(\alpha + \omega)})\}, \qquad \beta = \rho_d / \rho.$$

Equations (21) to (26) give nontrivial solutions if the determinant of the below matrix is zero –

$$\begin{bmatrix} iHk\cos\theta & 0 & 0 & -(\omega+\eta k^2) & 0 & 0 \\ 0 & iHk\cos\theta & 0 & 0 & -(\omega+\eta k^2) & 0 \\ -iBk\sin\theta & 0 & 0 & 0 & 0 & -(\omega+\eta k^2) \\ \omega(1+\frac{\beta v_c}{\omega+v_c})+D & 0 & s & -\frac{iBk\cos\theta}{4\pi\rho} & 0 & \frac{iBk\sin\theta}{4\pi\rho} \\ 0 & \omega(1+\frac{\beta v_c}{\omega+v_c})+\nu k^2 & 0 & 0 & -\frac{iVk\cos\theta}{4\pi\rho} & 0 \\ s & 0 & \omega(1+\frac{\beta v_c}{\omega+v_c})+J & 0 & 0 & 0 \end{bmatrix} \begin{bmatrix} u_x \\ u_y \\ u_z \\ h_x \\ h_y \\ h_z \end{bmatrix} = 0$$

On solving the determinant, we get

$$(\omega+\eta k^2)(\omega+\frac{\beta\omega v_c}{\omega+v_c}+J)V_a^4k^4\cos^4\theta+(\omega+\eta k^2)^2(\omega+\frac{\beta\omega v_c}{\omega+v_c}+vk^2)$$

$$\times(\omega+\frac{\beta\omega v_c}{\omega+v_c}+J)V_a^2k^2\cos^2\theta+(\omega+\eta k^2)V_a^4k^4\sin^2\theta\cos^2\theta(\omega+\frac{\beta\omega v_c}{\omega+v_c}+J)$$

$$+(\omega+\eta k^2)^2(\omega+\frac{\beta\omega v_c}{\omega+v_c}+D)(\omega+\frac{\beta\omega v_c}{\omega+v_c}+J)V_a^2k^2\cos^2\theta-(\omega+\eta k^2)^2V_a^2k^2\cos^2\theta s^2$$

$$+(\omega+\eta k^2)^2(\omega+\frac{\beta v_c\omega}{\omega+v_c}+vk^2)(\omega+\frac{\beta v_c\omega}{\omega+v_c}+J)V_a^2k^2\sin^2\theta+(\omega+\eta k^2)^3$$

$$\times(\omega+\frac{\beta v_c\omega}{\omega+v_c}+D)(\omega+\frac{\beta v_c\omega}{\omega+v_c}+vk^2)(\omega+\frac{\beta v_c\omega}{\omega+v_c}+J)-(\omega+\eta k^2)^3$$

$$\times(\omega+\frac{\beta v_c\omega}{\omega+v_c}+vk^2)s^2=0, \tag{28}$$

where Alfven velocity is given as $V_a = \dfrac{B}{(4\pi\rho)^{1/2}}$.

The above equation (28) provides the universal dispersion relation for an infinite homogeneous, magnetized, viscous, radiating, thermally and electrically conducting, partially-ionized plasma with neutral collision. The above general dispersion relation can be established with additional former result acquired. In nonexistence of thermal conductivity and radiative heat-loss function the dispersion relation (28) is matching to Bhatia & Hazarika [29] for non-rotational and non-Hall case. If we disregard the effect of radiative heat-loss function the dispersion relation (28) is matching to Shaikh et al. [30].

5. Discussion

To argue the consequences more successfully this general dispersion relation is discussed for transverse wave propagation.

5.1. Transverse Propagation ($\theta = 90°$)

In this case perturbations are taken perpendicular to the direction of magnetic field, i.e., $k\,sin\theta = k,\ k\,cos\theta = 0$, general dispersion relation (28) reduces to

$$(\omega+\eta k^2)^2(\omega+\frac{\omega\beta v_c}{\omega+v_c}+vk^2)^2[(\omega+\eta k^2)\{\omega+\frac{\omega\beta v_c}{\omega+v_c}+\frac{4}{3}vk^2 + \frac{(\Omega_I^2+\omega\Omega_J^2)}{\omega(\alpha+\omega)}\}+V_a^2k^2]=0. \tag{29}$$

This equation has three dissimilar features, we argue them separately. On substituting first aspect equal to zero, we get

$$(\omega+\eta k^2)=0. \tag{30}$$

This is represents a damped stable mode modified by finite electrical resistivity of the medium.

On replacement the second feature of scattering relation (29) equal to zero, we get

$$\omega^2 + \omega\{v_c(1+\beta) + vk^2\} + vk^2 v_c = 0\,. \tag{31}$$

This symbolizes a viscous type of clammy stable method customized by the being there of viscosity, neutral collision frequency and density ratio of neutral to ionized component. The above dispersion relation (31) is sovereign of thermal conductivity, radiative heat-loss function, magnetic field strength and finite electrical resistivity.

On substituting the third feature of dispersion relation (29) equivalent to zero, we get

$$\omega^5 + \omega^4\{B + v_c(1+\beta) + \frac{4}{3}vk^2 + \eta k^2\} + \omega^3\{v_c(1+\beta)(\eta k^2 + \alpha) + \alpha(\frac{4}{3}vk^2 + \eta k^2) + \frac{4}{3}vk^2(v_c + \eta k^2) + V_a^2 k^2 + \Omega_J^2\} + \omega^2[v_c\{\frac{4}{3}vk^2\alpha + \eta k^2\alpha(1+\beta) + \Omega_J^2 + V_a^2 k^2 + \frac{4}{3}v\eta k^4\} + \Omega_I^2 + \eta k^2(\Omega_J^2 + \frac{4}{3}v\eta k^2\alpha) + \alpha V_a^2 k^2] + \omega[v_c\{\Omega_I^2 + \eta k^2(\Omega_J^2 + \frac{4}{3}vk^2\alpha) + \alpha V_a^2 k^2\} + \eta k^2\Omega_I^2] + v_c\eta k^2\Omega_I^2 = 0. \tag{32}$$

The above equation represents the combined influence of all the considered physical parameters on the growth rate of thermal instability. To discuss the effect of each physical parameter on the growth rate of the thermal instability, we discussed the dispersion relation (32) in different cases.

Now taking the collective effect of all the parameters in the dispersion relation (32) of near difficulty, the modified thermal condition of instability obtained from the constant term of dispersion relation (32) is given as

$$\Omega_I^2 = \{k^2(T\zeta_T - \rho\zeta_\rho + \frac{Kk^2T}{\rho})\} < 0. \quad (33)$$

This is the modified thermal criterion of instability owing to thermal conductive and radiative effects. It is identical with equation (32). To see the effect of different physical parameters on the growth rate of instability, the dispersion relation (32) may be evaluated numerically in a simple manner. The value of γ in numerical calculations is taken as 5/3. We introduce the dimensionless parameters.

We solve equation (32) numerically by introducing the following dimensionless quantities

$$\omega^* = \frac{\omega}{k_\rho c_s}, \quad \nu^* = \frac{\nu k_\rho}{c_s}, \quad k^* = \frac{k}{k_\rho}, \quad k_\lambda^* = \frac{k_\rho}{k_\lambda}, \quad k_T^* = \frac{k_T}{k_\rho},$$
$$\eta^* = \frac{\eta k_\rho}{c_s}, \quad \nu_0{}^* = \frac{\nu_0 k_\rho}{c_s}. \quad (34)$$

Using Eq. (34), we write Eq. (32) in non-dimensional form as

$$\omega^{*5} + \omega^{*4}\{k_T^* + k_\lambda^* k^{*2} + v_c^*(1+\beta) + \frac{4}{3}v^* k^{*2} + \eta^* k^{*2}\} + \omega^{*3}\{v_c^*(1+\beta)(\eta^* k^{*2}$$
$$+ k_T^* + k_\lambda^* k^{*2}) + \left(k_T^* + k_\lambda^* k^{*2}\right)(\frac{4}{3}v^* k^{*2} + \eta^* k^{*2}) + \frac{4}{3}v^* k^{*2}(v_c^* + \eta^* k^{*2}) + V_a^{*2}k^{*2} + c^* k^{*2}\}$$
$$+ \omega^{*2}[v_c^*\{\frac{4}{3}v^* k^{*2}\left(k_T^* + k_\lambda^* k^{*2}\right) + \eta^* k^{*2}(k_T^* + k_\lambda^* k^{*2})(1+\beta) + c^* k^{*2} + V_a^{*2}k^{*2} + \frac{4}{3}v^* k^{*4}\}$$
$$+ \frac{k^{*2}}{\gamma}\left[k_T^* + k_\lambda^* k^{*2} - 1\right] + \eta^* k^{*2}(c^* k^{*2} + + \frac{4}{3}v^* \eta^* k^{*4} k_T^* + k_\lambda^* k^{*2}) + (k_T^* + k_\lambda^* k^{*2})V_a^{*2}k^{*2}]$$

$$+\omega^*[v_c^*\{\frac{k^{*2}}{\gamma}\left[k_T^*+k_\lambda^* k^{*2}-1\right]+\eta^* k^{*2}[c^* k^{*2}+\frac{4}{3}v^* k^{*2}(k_T^*+k_\lambda^* k^{*2})]+(k_T^*+k_\lambda^* k^{*2})V_a^2 k^2\}$$

$$+\eta^* k^{*2}\frac{k^{*2}}{\gamma}\left\{k_T^*+k_\lambda^* k^{*2}-1\right\}\Big]+v_c^*\eta^* k^{*2}\frac{k^{*2}}{\gamma}\left[k_T^*+k_\lambda^* k^{*2}-1\right]=0. \quad (35)$$

In Figures 1-6 the dimensionless growth rate (ω^*) has been plotted against the dimensionless wave number (k^*) to see the effect of various physical parameters such as radiative heat-loss function, resistivity and collision frequency. From Figure 1 we see that as the value of k_λ^* increases the growth rate decreases. Thus the effect of parameter k_λ^* is also stabilizing. From Figure 2 we conclude that growth rate decreases with increasing parameter k_T^*. Thus the presence of k_T^* stabilizes the growth rate of the system. It is clear from Figure 3 that growth rate decreases with increasing the value of viscosity. Thus the effect of viscosity is also stabilizing. Figure 4 shows the effect of resistivity on the growth rate of thermal instability. From curves it is clear that the growth rate of thermal instability increases as the value of resistivity increases. Hence the resistivity has destabilizing influence on the system. Figure 5 displays the influence of collision frequency on the growth rate of thermal instability. From figure it is clear that the collision frequency has a destabilizing effect on the growth rate of thermal instability. It is clear from Figure 11 that growth rate increases with increasing the value of β. Thus the effect of β is destabilizing. Therefore, the parameters radiative heat-loss functions and viscosity have stabilizing influence on the system while the resistivity, collision frequency and β have destabilizing influence on the growth rate of the system.

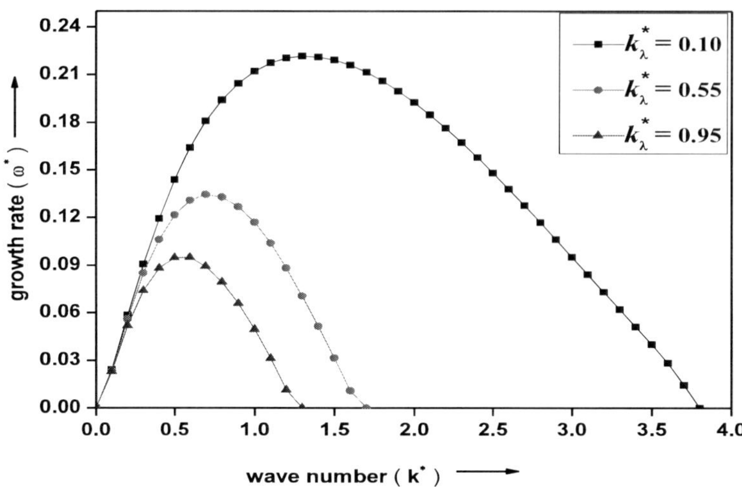

Figure 1. The normalized growth rate (ω^*) as a function of normalized wave number (k^*) for different values of k_λ^* having $\eta^* = 1$ with $k_T^* = 0.5$and $\nu^* = \nu_c^* = \beta = 1$.

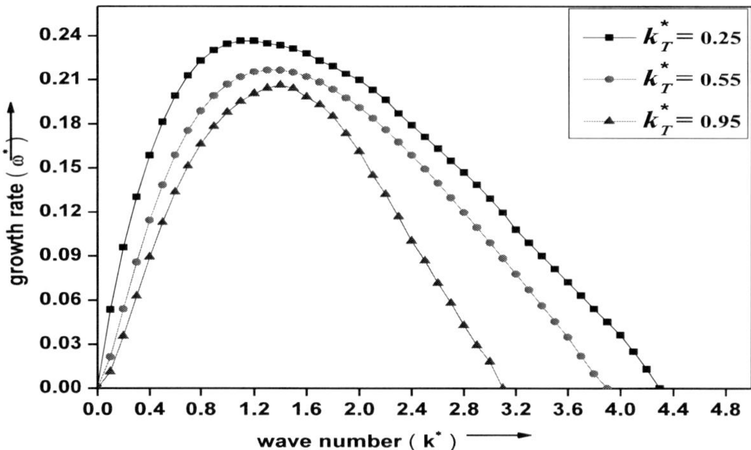

Figure 2. The normalized growth rate (ω^*) as a function of normalized wave number (k^*) for different values of k_T^* having $\eta^* = 1$ with $k_\lambda^* = 1$ and $\nu^* = \nu_c^* = \beta = 1$.

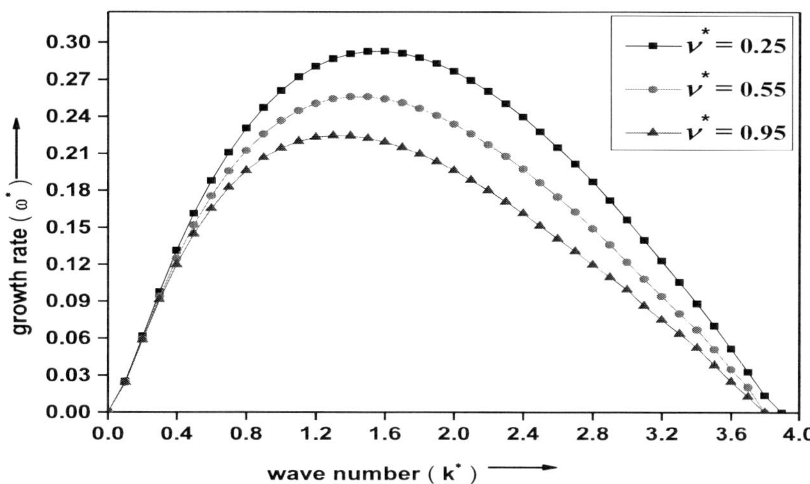

Figure 3. The normalized growth rate (ω^*) as a function of normalized wave number (k^*) for different values of v^* having $k_T^* = 0.5$ with k_λ^* = 1 and $\eta^* = v_c^* = \beta = 1$.

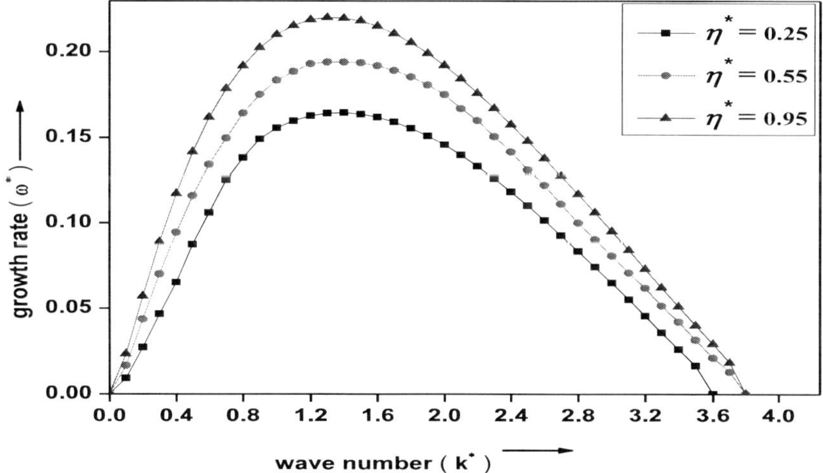

Figure 4. The normalized growth rate (ω^*) as a function of normalized wave number (k^*) for different values of η^* having $k_\lambda^* = .1$ with k_T^* = 0.5and $v^* = v_c^* = \beta = 1$.

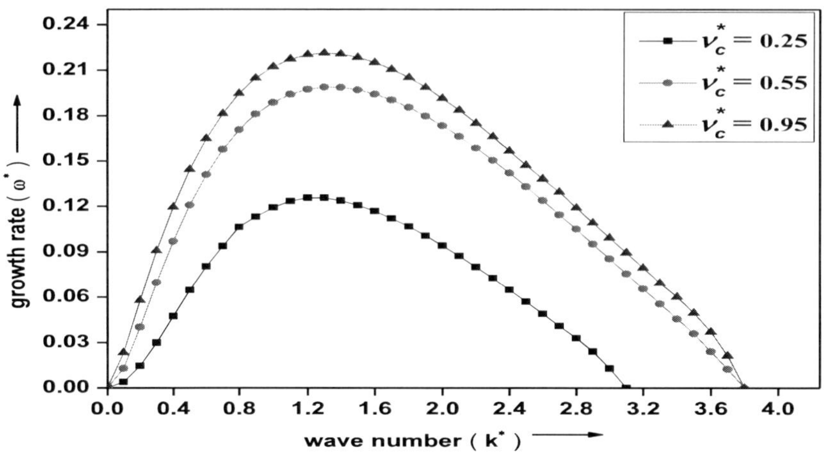

Figure 5. The normalized growth rate (ω^*) as a function of normalized wave number (k^*) for different values of v_c^* having $k_\lambda^* = .1$ with $k_T^* = 0.5$ and $v^* = \eta^* = \beta = 1$.

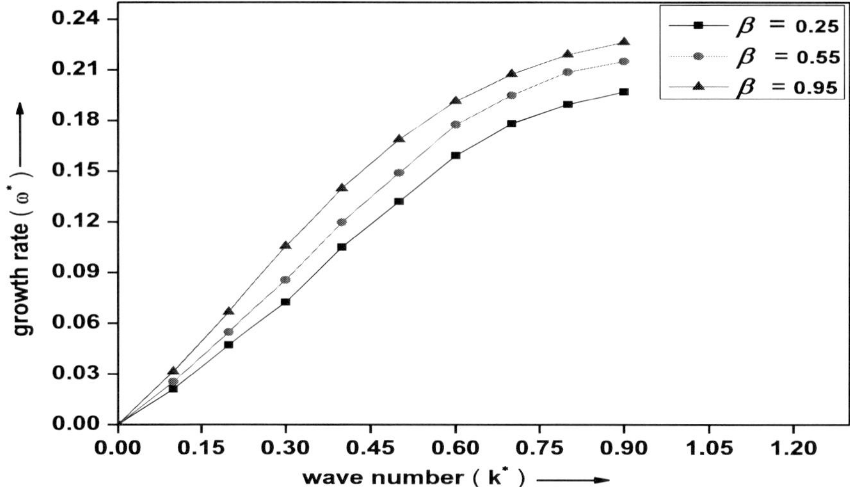

Figure 6. The normalized growth rate (ω^*) as a function of normalized wave number (k^*) for different values of β having $k_\lambda^* = .1$ with $k_T^* = 0.5$ and $v^* = \eta^* = v_c^* = 1$.

From the above discussion we conclude that thermal criterion of instability is modified due to magnetic field, finite electrical resistivity, thermal conductivity, and radiative effects in the transverse mode of propagation. The effect of collisions with neutrals and the density ratio of neutral to ionized component do not affect the condition of instability in this case, but presence of these parameters modify the growth rate of instability.

CONCLUSION

We have analyzed the effect of neutral collision and radiative heat-loss function on the thermal instability of partially-ionized plasma with thermal conductivity, viscosity and finite electrical resistivity. We locate that, viscosity; finite electrical resistivity and collision frequency of the ionized and neutral components give stable mode. The presence of collisions with neutrals does not affect the instability condition in all the cases. It is found that collisions with neutrals, viscosity and finite electrical resistivity influence the longitudinal mode of wave propagation by modifying the Alfven mode. We conclude that the thermal conductivity modifies the thermal condition of instability by changing adiabatic sound velocity with isothermal sound velocity. The thermal length is reduced due to thermal conductivity. Owing to the inclusion of radiative effects the critical thermal wave number is very much different from the classical thermal value and depends on heat-loss function in a complicated way. The finite electrical resistivity removes the effect of magnetic field from the thermal condition of instability assuming the system to behave as unmagnetized in transverse mode of wave propagation.

To see the effect of various physical parameters on the growth rate of unstable mode numerical calculations have been performed. The heat-loss function, thermal conductivity, neutral collision frequency, viscosity and magnetic field have stabilizing influence while finite electrical resistivity has a destabilizing influence on the growth rate of thermal instability.

From the above discussed outcomes we express to a lock that, when the cloud density reaches your destination risky importance, the cloud fragments into chill dense condensations via thermal instability. When the serious density enlarges as metallicity diminishes, and also as radiation expands. Condensations collide with each other and self-gravitating clumps will be fashioned when the indicate cloud density becomes satisfactorily elevated; then stars will form. Additional room of the H II region in the state of the enormous star and supernova explosions will blow off neighboring gas and end star configuration procedure. When the require density at the time of star configuration is elevated, immense virial velocity stops development of the H II region. Also, in such elevated-density surroundings, the star structure timescale is shorter than the lifetime of a enormous star. Then the gas in cluster-appearance section will be rehabilitated into stars resourcefully, previous to the gas is standing apart by mounting H II region or supernova detonations. So happen density is realize in the contracting near to the ground-metallicity gas, and if the configuration of a gathering gas cloud is probable, a physically powerful-radiation surroundings is an additional contender. Thus, it is recommend that towering star configuration good organization and jump cluster structure are look onward to accomplish in near to the ground-metallicity and/or physically powerful-radiation situations. Such surroundings continue living in dwarf galaxies, the in the early hours point in time of our galaxy and starburst galaxies.

REFERENCES

[1] Field, G. B., (1965). Thermal instability, *Astrophys. J.*, 142:531.

[2] Hunter, J. H., (1966). Thermal instability in the solar chromosphere, *Icaras* 5:321.

[3] Cook, J. W., Cheng, C. C., Jacobs, V. L. & Antiochos, S. K., (1988). Effects of coronal elemental abundances on the radiative loss function *Astrophys. J.*, 338:1176.

[4] Van hoven, G. & Mok, V., (1984). The thermal instability in a sheared magnetic field: Filamental condensation with anisotropic heat conduction, *Astrophys. J.*, 282:267.

[5] Hunter, J. H., (1966). The role of thermal instability in star formation, *Mon. Not. R. Astron. Soc.*, 133:239.

[6] Aggarwal, M. & Talwar, S. P., (1969). Magnetothermal instability in arotating gravitationg fluid, *Mon. Not. R. Astron. Soc.*, 146:235.

[7] Aggarwal, M. & Talwar, S. P., (1969). Constreaming instability in gravitating fluids with thermal effects, *Publ. Astron. soc. Japan* 21:176.

[8] Beltrametti, M., (1981). Thermal instability in a radiatively driven winds-Application to emission line clouds for quasar and active galactic nuclei, *Astrophys. J.*, 250:18.

[9] Gupta, M. R., Tanuka, K. & Basu, B., (1991). Thermal instability analysis of the growth f molecular clouds in our galaxy, *Astrophys. Space Sci.*, 176:85.

[10] Ibanez, S. M. H., (1985). Sound and thermal waves in a fluid with an arbitrary heat-loss function, *Astrophys. J.*, 290:33.

[11] Kim, W. & Narayan, R., (2003). Thermal instability in clusters of galaxies with conduction, *Astrophys. J.*, 596:889.

[12] Radwan, A. E., (2004). Variable streams self-gravitating instability of radiative rotating gas cloud, *App. Mats and Comp.* 148:331.

[13] Menou, K., Balbus, A. S. & Spruit, C. H., (2004). Local axiaymmetric Diffusive stability of weakly magnetized, differentialy rotating, stratified fluids, *Astrophys. J.*, 607:564.

[14] Inutsuka, S., Koyama, H. & Inoue, T. (2005), Magnetic Fields in the Universe: From Laboratory and Stars to Primordial

Structures,ed. E. M. de Gouveia dal Pino, G. Lugones, & A. Lazarian (Melville: AIP), *AIP Conf. Proc.*, 784:381.

[15] Shadmehri, M. & Dib, S., (2009). Magnetothermal condensation modes including the effects of charged dust particles, *Mon. Not. R. Astron. Soc.*, 395:985.

[16] Bora, M. P. & Talwar, S. P., (1993). Magnetothermal instability with generalized Ohm's law *Phys. Fluids.*, *B* 5:950.

[17] Renard, M. & Chieze, J. P., (1993). The fragmentation of molecular clouds: critical (Jeans) mass in the vicinity of thermal instability and influence of visible extinction vibrations, *Astron. Astrophys.*, 267:549.

[18] Dwivedi, C. B., Singh, R. & Avinash, K., (1996). Effect of radiative condensation on Jeans instability, *Phys.*, *Scr.*, 53:760.

[19] Talwar, S. P. & Bora, M. P., (1995). Thermal instability in a star gas system, *J. Plasma Phys.*, 54:157.

[20] Prajapati, R. P., Pensia, R. K., Kaothekar, S. & Chhajlani, R. K., (2010). Self-gratitational instability of rotating viscous Hall plasma with arbitrary radiative heat-loss functions and electron inertia, *Astrophys. Space Sci.*, 327:139.

[21] Hobbs, A., Read, J., Power, C. & Cole, D., (2012). Thermal instability in cooling galactic coronae: fuelling star forming in galactic disc, *Mon. Not. R. Astron. Soc.*, 434:2300.

[22] Dutta, J., (2015). On the effects of rotation in primordial star forming clouds, *Astrophys. J.*, 811:6.

[23] Ling, G. & Thuns, T. & Volker, S., (2015). Star forming filaments in warm dark matter models, *Mon. Not. R. Astron. Soc.*, 450:45.

[24] Khesali, A. R., Ghoreyshi, S. M. & Najad-asfhar, M., (2012). Thermal instability in a molecular clouds, including dust particles, Hall effect and ambipolar diffusion, *Mon. Not. R. Astron. Soc.* 420, 2300.

[25] Voit, G. M., Donahue, M., Brayn, G. L. & Mcdonal, M. (2015). Regulation of star formation in giant galaxies by precipitation, feedback and conduction, *Nature*, 519:203.

[26] Wareing, C. J., Pittard, J. M. & Falle, S. A. E. G. (2016). Sheets, filaments and clumps-high-resolution simulations of how the thermal instability can form molecular clouds, *Mon. Not. R. Astron. Soc.* 465:2757.

[27] McNamara, B. R., Russell, H. R., Nulsen, P. E. J., Hogan, M. T., Fabian, A. C. et al., (2014). A 10^{10} solar mass flow of molecular gas in the A1835 brightest clusture gasaxy, *Astrophy. J.*, 785:45M.

[28] Ali, A. & Bhatia, P. K., (1993). Gravitational instability of thermaly conducting viscous Hall plasma, *Phys. Scr.*, 47:561.

[29] Bhatia, P. K. & Rajib hazarika, A. B., (1995). Gravitational instability of partially-ionized plasma in an oblique manetic field, *Phys. Scr.,* 51:775.

[30] Shaikh, S., Khan, A. & Bhatia, P. K., (2007). Thermally conducting partially-ionized plasma in variable magnetic field, *Contrib. Plasma Phys.*, 47:147.

[31] Jain, S., Sharma, P., Kaothekar, S. & Chhajlani, R. K., (2016). Effect of finite Larmor radius correcctions on the thermal instability of thermally conducting viscous plasma with Hall current and electron inertia, *Astrophys. J.*, 829:122.

[32] Kaothekar, S., (2017). Molecular cloud formation via thermal instability of finite resistive viscous radiating plasma with finite Larmor radisu correction, *Astrophys. Space Sci.,* 362:107.

[33] Kaothekar, S., (2018). Thermal instability of partially ionized viscous plasma with Hall effect FLR corrections flowing through porous medis, *J. Porous Media* 21, 679.

In: An Introduction to Molecular Clouds ISBN: 978-1-53619-178-3
Editor: Sachin Kaothekar

Chapter 6

INFLUENCE OF ELECTRON PLASMA FREQUENCY ON THE EVOLUTION OF GRAVITATIONAL MOLECULAR CLOUDS

D. L. Sutar[1,*], R. K. Pensia[2], A. K. Patidar[3] and V. Shrivastva[4]

[1]Department of Physics, Government RV College
Manasa (MP), India
[2]Department of Physics, Government Girls PG College,
Neemuch (MP), India
[3]Department of Physics, Government PG College,
Mandsaur (MP), India
[4]Department of Physics, Government Art and Science College,
Ratlam (MP), India

---

* Corresponding Author's Email: devilalsutar833@gmail.com.

ABSTRACT

Throughout this chapter, we have examined the effect of electron plasma frequency on gravitational instability in the magneto-radiation quantum plasma region. The research is carried out within the context of the normal mode analysis method, which has been improved due to the involvement of the electron plasma frequency. The dispersion relationship is simplified in the longitudinal and transverse propagation of the magnetic field. The expression instability is modified due to the existence of the magnetic field and the electron plasma frequency in the transverse mode of propagation. It is obvious from the curve that the electron plasma frequency has a destabilizing effect on the system. However, the presence of a quantum parameter reduces the detrimental effect of electron plasma frequency and stabilizes the system. The importance of our research will allow us to better understand the stellar evolution of self-gravitational molecular clouds and the creation of stars.

Keywords: quantum correction, electron plasma frequency, electrical resistivity, radiative heat-loss function, and magnetic field

1. INTRODUCTION

Jeans instability is the keywords for explaining the mechanism of star-forming that Jeans has discovered [1] that it plays a central role in the creation of stellar objects, i.e., stars are created as scattered matter in gaseous form starts to coalesce under the force of gravity, constantly attempting to shrink the objects and in this process, the density and temperature rising within the objects. The period of a billion years is required for the creation of celestial objects, while Jeans calculates that the process is comparatively quicker. According to Jean's criterion, the infinite homogeneous self-gravitational atmosphere is unstable for all wave-numbers k less than Jean's wave-number $k_j = \left(\frac{G\rho}{S}\right)$ where ρ is the density, S is the velocity of sound in the gas, and G is the gravitational constant. Several scholars have researched this problem under various hypotheses of hydrodynamics and hydromagnetics, and Chandrasekhar [2] provided a detailed account of these investigations in his monograph on the problems of hydrodynamics and hydromagnetic stability. He observed that the criteria of Jeans were unchanged by the independent or overlapping presence of rotation and magnetic field. Gomez-Pelaez and Moreno-Insertis [3] have studied the thermal instability in the cooling and expansion medium, including self-gravity and conduction in neutral fluid dynamics. The thermal equilibrium of spinning optically thin plasma in a weakly magnetic field has been studied by Nipoti and Posti [4]. The stability properties of thermal modes in cool plasma prominence were studied by Soler et al., [5]. Hobbs et al., [6] explored thermal instability in a cooling galactic corona fueling star formation in a galactic disc. Inoue and Omukai [7] were responsible for the issue of thermal instability and the multi-phase interstellar medium in the first galaxies. It is well known that the quantum effect plays a significant role in the creation of structures by the gravitational collapse of astrophysical objects. Pines [8] first introduced quantum plasma, he studied that at very low temperature, the de Broglie wavelength

$$\lambda_B = \frac{h}{\sqrt{2m_e, K_b T}} \text{(where } m_e, K_b, T \text{)}$$

are the electron and ion masses, Boltzmann constant, and temperature of electrons and ions is of the order of the dimension of the system, such as Debye length and Larmor radius. In this sort of dense plasma system, the wave function is correlated with overlapping particles, so the plasma behaves like Fermi gas, and we will regard it as quantum plasmas. The quantum plasma function can be described by three popular models, the Schrodinger-Poisson (SP) model or the quantum hydrodynamic (QHD) model, the Wigner-Poisson (WP) model, and the Hartree model. Hass [9] provides a quantum multi-stream model for single-and two-stream plasma instability and also explores the stationary status of the Schrodinger-Poisson non-linear model. Masood et al., [10] examined the self-gravitational instability of multi-component quantum plasma using the potential of Bohm and the statistical parameters of electrons and ions. In this way, several authors [11-14] explored the uncertainty of Jean, including the various criteria of their analysis. Sharma et al., [15] recently studied the influence of electron plasma frequency on the radiative instability of optically dense plasma. Kumar et al., [16] addressed the influence of photoelectron current on jeans instability of spinning quantum dusty plasma.

As a result, a vast number of experiments are performed on the quantum magnetohydrodynamic model (QMHD) with different parameters under different assumptions. Yet no one is contemplating the quantum a magnetohydrodynamic model with resistance and electron plasma frequency effects.

2. The Problematic Equation

In this research, we assume an infinite self-gravitational homogeneous plasma medium, composed of electrons and single charged ions with electrical resistance. Plasma is immersed in a uniform magnetic field of the ambient kind $\vec{H}(0,0,\mathrm{H})$ in the z-direction. The fundamental radiative QMHD series of equations follows:

$$\frac{\partial \vec{V}}{\partial t} = -\frac{\nabla \delta p}{\rho} + \nabla \delta U + \frac{1}{4\pi\rho}\left(\nabla \times \vec{h}\right) \times \vec{H} + \frac{\hbar^2}{4m_e m_i}\nabla \frac{\left(\nabla^2 \delta\rho\right)}{\rho} \tag{1}$$

$$\frac{\partial \delta\rho}{\partial t} = -\rho \nabla . \vec{V} \tag{2}$$

$$\nabla^2 \partial U = -4\pi G \delta\rho \tag{3}$$

$$\frac{1}{(\gamma-1)}\frac{\partial \delta p}{\partial t} - \frac{\gamma}{(\gamma-1)}\frac{p}{\rho}\frac{\partial \delta\rho}{\partial t} + \rho\left(\mathcal{L}_\rho \delta\rho + \mathcal{L}_T \delta T\right) = \lambda \nabla^2 \delta T \tag{4}$$

$$\frac{\delta P}{P} = \frac{\delta T}{T} + \frac{\delta \rho}{\rho} \tag{5}$$

$$\frac{\partial \vec{h}}{\partial t} = \nabla \times \left(\vec{V} \times \vec{H}\right) + \eta \nabla^2 \vec{h} + \frac{c^2}{\omega_{pe}^2} \frac{\partial}{\partial t} \nabla^2 \vec{h} \tag{6}$$

$$\nabla . \vec{h} = 0 \tag{7}$$

The equations mentioned above (1) to (7) are momentum transfer equation, continuity equation, Poisson's equation, heat equation for a perfect gas and state equation, idealized Ohm's law with electron plasma frequency and resistivity, Gauss's law respectively.

Where, $\vec{V}(V_x, V_y, V_z)$, is the fluid velocity, δp is the fluid pressure, U gravitational potential, $\vec{h}(h_x, h_y, h_{z,})$ is the magnetic field, $\delta \rho$ is the fluid density, G gravitational constant, γ is the ratio of two specific heat, δT is the temperature, λ is the thermal conductivity, $\mathcal{L}_\rho$ is the partial derivatives of the density-dependent $(\partial \mathcal{L}/\partial T)_T$ heat-loss function, $\mathcal{L}_T$ is the partial derivatives of the temperature-dependent $(\partial \mathcal{L}/\partial T)_\rho$ heat-loss functions, δT is the temperature, $\hbar$ Plank's constant divided by $2\pi, m_e$ *and* m_i are the electron and ion mass, and ω_{pe} is the electron plasma frequency.

Combining equation (4) and (5), we obtain the expression for δp as

$$\delta p = \left(\frac{\alpha + \sigma C^2}{\sigma + \beta}\right) \delta \rho \tag{8}$$

where $\sigma = i\omega$ is the growth rate of the perturbation, and $C = \left(\frac{\gamma p}{\rho}\right)^{1/2}$ is the adiabatic velocity of sound in the medium, $s = \delta \rho / \rho$ is the condensation of the medium. The parameter α *and* β are

$$\alpha = (\gamma - 1)\left(\mathcal{L}_T T - \mathcal{L}_\rho \rho + \frac{\lambda k^2 T}{\rho}\right) \text{ and } \beta = (\gamma - 1)\left(\frac{\mathcal{L}_T T \rho}{p} + \frac{\lambda k^2 T}{p}\right)$$

We presume that all the perturbed quantities vary as

$$exp\{i(k_x x + k_z z + \omega t)\} \tag{9}$$

Where k_x and k_z are the wavenumbers in perpendicular and parallel direction to the magnetic field, such that $k_x^2 + k_z^2 = k^2$, ω is the frequency of harmonic disturbances respectively.

Using equation (1) - (9) we're getting the following matrix relation.

$$X_{ij} Y_j = 0, i, j = 1,2,3,4, \tag{10}$$

Where X_{ij} is a 4×4 matrix whose elements are,

$$X_{11} = \left(\sigma + \frac{k^2 V^2}{A_1}\right),\ X_{12} = 0\ X_{13} = 0, X_{14} = \frac{ik_x}{k^2}\left(\Omega_T^2 + \frac{\hbar^2 k^4}{4m_e m_i}\right)$$

$$X_{21} = 0, X_{22} = \left(\sigma + \frac{k_z^2 V^2}{A_1}\right), X_{23} = 0, X_{24} = 0,$$

$$X_{31} = 0, X_{32} = 0,\ X_{33} = \sigma, X_{34} = \frac{ik_z}{k^2}\left(\Omega_T^2 + \frac{\hbar^2 k^4}{4m_e m_i}\right),$$

$$X_{41} = \frac{ik_x k^2 V^2}{A_1}, X_{42} = 0, X_{43} = 0, X_{44} = -\left(\sigma^2 + \Omega_T^2 + \frac{\hbar^2 k^4}{4m_e m_i}\right)$$

Where $V = \frac{H}{(4\pi\rho)^{1/2}}$ is the Alfven velocity, $C' = \left(\frac{p}{\rho}\right)^{1/2}$ are isothermal velocities of sound, respectively. Also, we have assumed the following substitution.

$$\Omega_J^2 = (k^2 C^2 - 4\pi G\rho), \Omega_I^2 = (k^2 \alpha - 4\pi G\rho\beta), \Omega_T^2 = \left(\frac{\sigma\Omega_J^2 + \Omega_I^2}{\sigma + \beta}\right), \Omega_m$$
$$= \eta k^2, A_1 = (\sigma f + \Omega_m), f = \left(1 + \frac{c^2 k^2}{\omega_{pe}^2}\right),$$

The general dispersion relation can be obtained from the determinant of the matrix of equation (10) gives the general dispersion relation as

$$\sigma\left(\sigma+\frac{k^2V^2}{A_1}\right)\left(\sigma+\frac{k_z^2V^2}{A_1}\right)\left(\sigma^2+\Omega_T^2+\frac{\hbar^2k^4}{4m_em_i}\right)-\sigma\frac{k_x^2}{k^2}\left(\Omega_T^2+\frac{\hbar^2k^4}{4m_em_i}\right)\left(\frac{k^2V^2}{A_1}\right)\left(\sigma+\frac{k_z^2V^2}{A_1}\right)=0 \quad (11)$$

The dispersion relationship (11) illustrates the overall influence of electron plasma frequency, electrical resistance, radiative heat loss function, thermal conductivity, quantum correction, and gravitational mode of the system. If the influence of electron plasma frequency, resistance, quantum correction, and the magnetic field strength is ignored, the dispersion ratio (11) is close to Ibanez [11].

3. DISCUSSION

For the detailed investigation of the influence of electron plasma frequency and electrical resistivity including the radiative magnetized quantum plasma, the dispersion relation is discussed for parallel and perpendicular propagations.

3.1. Parallel Propagation

In that kind of parallel propagation, we assume that all disturbances are longitudinal to the direction of the magnetic field (k x=0, k z = k). Thus, the relation of dispersion (11) can be summarized as follows.

$$\sigma\left(\sigma+\frac{k^2V^2}{A_1}\right)^2\left(\sigma^2+\Omega_T^2+\frac{\hbar^2k^4}{4m_em_i}\right)=0 \quad (12)$$

The equation (12) indicates the simultaneous effect of thermal conductivity, magnetic field, self-gravitation, quantum plasma, electrical resistance, heat loss function, and electron plasma frequency, the equation above has tree-independent variables, each representing various parameters. The first factor of the equation (12) is $\sigma = 0$ and represents the natural stability of the system. The second factor of the equation (12) illustrates the second-order equation,

$$\sigma^2 f + \sigma\,\Omega_{\mathrm{m}} + k^2 V^2 = 0 \tag{13}$$

The dispersion relation (13) is affected by the presence of electron plasma frequency, resistivity, and magnetic field but is dispersion relation does not effect by heat loss function and quantum correction. Now the third factor of the equation of (12) is equating to zero and solved it we get the dispersion relation.

$$\sigma^3 + \sigma^2\beta + \sigma\left(\Omega_j^2 + \frac{\hbar^2 k^4}{4m_e m_i}\right) + \Omega_I^2 + \beta\frac{\hbar^2 k^4}{4m_e m_i} = 0 \tag{14}$$

The dispersion relation (14) is adjusted by thermal conductivity, quantum correction, and radiative heat loss function of the medium. This mode doesn't depend on the electron plasma frequency, resistivity, and magnetic field. The state of the volatility of Jeans is accomplished by a dispersion relation (14) given as

$$(\gamma - 1)\left(\mathcal{L}_T T - \mathcal{L}_\rho \rho + \frac{\lambda k^2 T}{\rho}\right) + \frac{\hbar^2 k^4}{4m_e m_i}\left(\frac{\mathcal{L}_T T\rho}{p} + \frac{\lambda k^2 T}{p}\right) < \frac{4\pi G\rho}{k^2} \tag{15}$$

The equation (15) expresses a modified state of Jeans instability due to quantum correction yet are independent of electrical resistance, magnetic field, and electron plasma frequency in the longitudinal mode of propagation.

3.2. Perpendicular Propagation

In this case, we assume that all the disturbances propagate perpendicular to the direction of the magnetic field, we take $k_x = k, k_z = 0$. The dispersion relation (11) may be written as

$$\sigma^4 f + \sigma^3(f\beta + \Omega_{\mathrm{m}}) + \sigma^2\left(f\Omega_j^2 + f\frac{\hbar^2 k^4}{4m_e m_i} + k^2 v^2 + \beta\Omega_{\mathrm{m}}\right) + \sigma\left(f\Omega_I^2 + \Omega_{\mathrm{m}}\Omega_j^2 + \frac{f\beta\hbar^2 k^4}{4m_e m_i} + \frac{\Omega_{\mathrm{m}}\hbar^2 k^4}{4m_e m_i} + \beta k^2 v^2\right) + \Omega_I^2\Omega_{\mathrm{m}} + \Omega_{\mathrm{m}}\beta\frac{\hbar^2 k^4}{4m_e m_i} = 0 \tag{16}$$

The dispersion relation (16) indicates the cumulative effect of the magnetic field, electron plasma frequency, quantum correction, thermal conductivity, radiative heat-loss functions, and gravitating mode of the system. Thus equation (16) represents a gravitating Alfven mode modified by quantum correction, electron plasma frequency, thermal conductivity, and radiative heat-loss functions. In the absence of quantum correction, radiative heat-loss function and thermal conductivity effect, this dispersion relation (16) is similar to obtained by Uberoi [12] excluding rotation effect in that case. If the quantum correction effect is ignored then (16), the result is equivalent to that of Bora and Talwar [13].

The state of instability is achieved via the constant term of the dispersion relation (16) given as

$$\left[\eta k^2(\gamma - 1)\left\{\left(\mathcal{L}_T T - \mathcal{L}_\rho \rho + \frac{\lambda k^2 T}{\rho}\right) + \frac{\hbar^2 k^4}{4m_e m_i}\left(\frac{\mathcal{L}_T T\rho}{p} + \frac{\lambda k^2 T}{p}\right)\right\} < \frac{4\pi G\rho}{k^2}\right] \tag{17}$$

Write the dispersion relation (16) in the dimensionless form to demonstrate the influence of various parameters on the growth rate of instability.

$$\sigma^{*4}f + \sigma^{*3}(f\beta^* + \eta^* k^{*2}) + \sigma^{*2}\{f(k^{*2} - 1) + fQ^* k^{*2} + k^{*2}V^{*2} + \beta^*\eta^* k^{*2}\} + \sigma^*\{f(k^{*2}\alpha^* - \beta^*) + \eta^* k^{*2}(k^{*2} - 1) + f\beta^* Q^* k^{*2} + \eta^* k^{*4}Q^* + \beta^* k^{*2}V^{*2}\} + \eta^* k^{*2}(k^{*2}\alpha^* - \beta^*) + \beta^*\eta^* k^{*4}Q^* = 0 \quad (18)$$

Where the various non-dimensional parameters are defined as

$$\sigma^* = \frac{\sigma}{\sqrt{4\pi G\rho}},\ k^* = \frac{kC}{\sqrt{4\pi G\rho}}, Q^* = \frac{\hbar^2 k_j^2}{4m_e m_i}, V^* = \frac{V\sqrt{4\pi G\rho}}{C}, \lambda^* = \frac{(\gamma-1)T\lambda\sqrt{4\pi G\rho}}{\rho C^2}, \mathcal{L}_\rho^* = \frac{(\gamma-1)\rho\mathcal{L}_\rho}{C^2\sqrt{4\pi G\rho}}, \mathcal{L}_T^* = \frac{(\gamma-1)\rho T\mathcal{L}_T}{\rho\sqrt{4\pi G\rho}}, \alpha^* = \left(\frac{1}{\gamma}(\mathcal{L}_T^* + \lambda^* k^{*2}) - \mathcal{L}_\rho^*\right), \beta^* = (\mathcal{L}_T^* + \lambda^* k^{*2}), \eta^* = \frac{\eta\sqrt{4\pi G\rho}}{C^2}, \Omega_I^{*2} = (k^{*2}\alpha^* - \beta^*), \Omega_j^{*2} = (k^{*2} - 1) \quad (19)$$

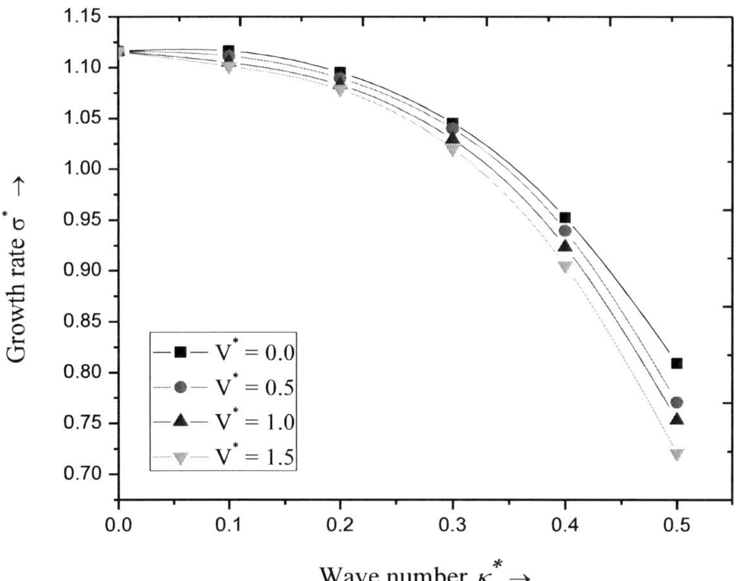

Figure 1. The growth rate (σ^*) is plotted against the non-dimensional wave number (k^*) with variation in the magnetic field $V^* = (0, 0.5, 1, 1.5)$, keeping the values of other parameters are fixed, as $\mathcal{L}_\rho^* = \eta^* = \mathcal{L}_T^* = \lambda^* = Q^* = 0.5\ and\ f = 0$.

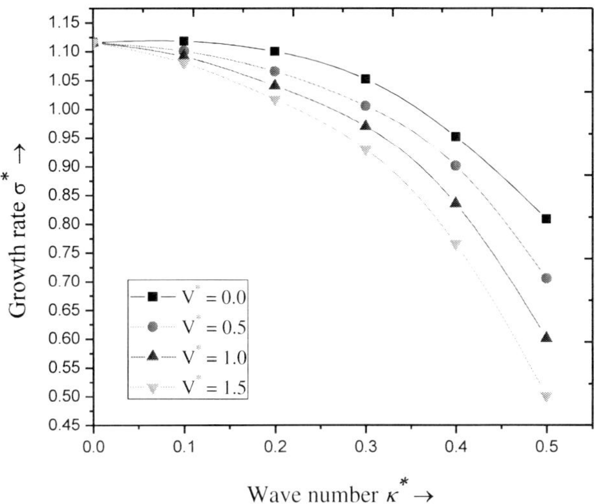

Figure 2. The growth rate (σ^*) is plotted against the non-dimensional wave number (k^*) with variation in the magnetic field $V^* = (0, 0.5, 1, 1.5)$, keeping the values of other parameters are fixed, as $\mathcal{L}_\rho^* = \eta^* = \mathcal{L}_T^* = \lambda^* = Q^* = 0.5$ *and* $f = 0.5$.

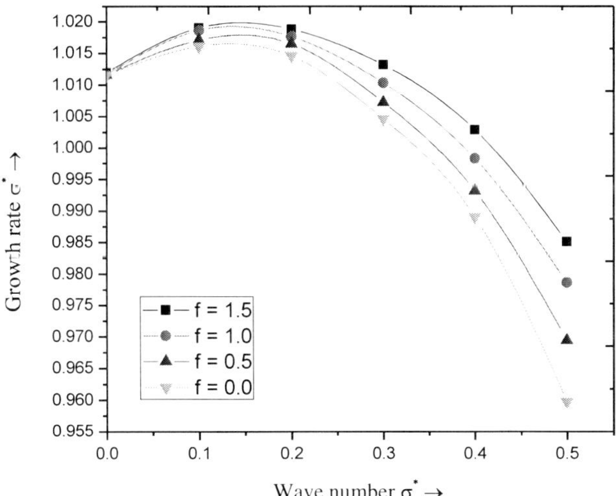

Figure 3. The growth rate (σ^*) is plotted against the non-dimensional wave number (k^*) with variation in the electron plasma frequency $\mathrm{f} = (0, 0.5, 1, 1.5)$, keeping the values of other parameters are fixed, as $\mathcal{L}_\rho^* = \mathcal{L}_T^* = \eta^* = \lambda^* = V^* = 0.5$ *and* $Q^* = 0$.

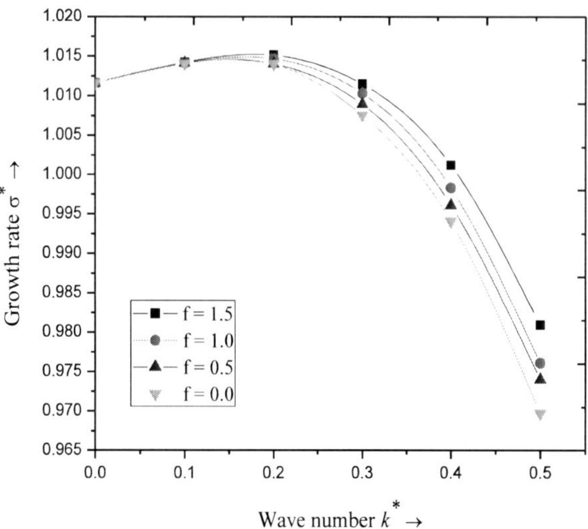

Figure 4. The growth rate (σ^*) is plotted against the non-dimensional wave number (k^*) with variation in the electron plasma frequency f = (0, 0.5, 1, 1.5), keeping the values of other parameters are fixed, as $\mathcal{L}_\rho^* = \mathcal{L}_T^* = \eta^* = \lambda^* = Q^* = 0.5$ *and* $Q^* = 0.5$.

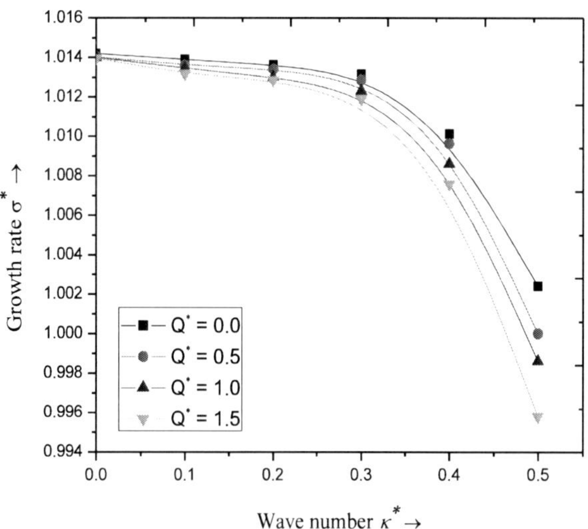

Figure 5. The growth rate (σ^*) is plotted against the non-dimensional wave number (k^*) with variation in the quantum correction $Q^* = (0, 0.5, 1, 1.5)$, keeping the values of other parameters are fixed, as $\mathcal{L}_\rho^* = \mathcal{L}_T^* = \lambda^* = \eta^* = V^* = 0.5$ *and* $f = 0$.

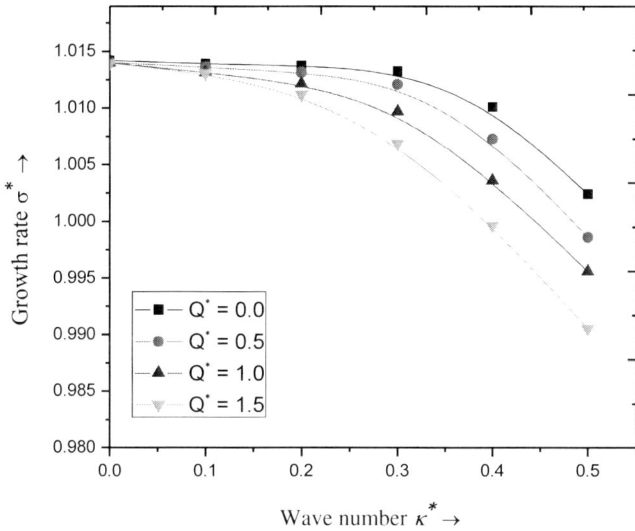

Figure 6. The growth rate (σ^*) is plotted against the non-dimensional wave number (k^*) with variation in the quantum correction $Q^* = (0, 0.5, 1, 1.5)$, keeping the values of other parameters are fixed, as $\mathcal{L}_\rho^* = \mathcal{L}_T^* = \lambda^* = \eta^* = V^* = 0.5\ and\ f = 0.5$.

From the curves, we find that in figure 1-2, the magnetic field has a stabilizing impact in the system but the appearance of electron plasma frequency the system is unstable. In figure 3-4 we conclude that the electron plasma frequency has a destabilizing effect but in the presence of quantum parameter the destabilizing effect is reduced. In the curve 5-6, we see that the quantum parameter has to stabilize effect but in the presence of electron plasma frequency the system is destabilized so that the quantum parameter tries to decrease the destabilizing effect.

CONCLUSION

The present dilemma was studied in the sense of the theory of quantum fluids. We obtained a general dispersion relation using a normal mode analysis and a QMHD equation. Throughout the case of parallel propagation, the Jeans instability is modified in the presence of radiation and quantum correction but is independent of the electron plasma frequency and electrical resistance. The perpendicular mode of the dispersion relation is influenced by both parameters. Gravitational instability is caused by electrical resistance, thermal conductivity, radiative heat loss mechanism, and quantum correction. In the graphical presentation, the result shows that the magnetic field and the quantum correction have stabilized the growth rate of the system, but the electron plasma frequency has a destabilizing effect on the growth rate of instability.

REFERENCES

[1] Jeans, J. H. The stability of a spherical nebula. *Phil. Trans. Roy. Soc*. London. 1902, 199, 1.

[2] Chandrashekhar, S. *Hydrodynamics and Hydromagnetic Stability* (Clarendon Press, Oxford, 1961).

[3] Gomez-Pelaez, A. J.; Moreno- Insertis, F. Thermal Instability in a Cooling and Expanding Medium Including Self-Gravity and Conduction. *The Astrophysical Journal*. 2002, 569(2), 766.

[4] Nipoti, C.; Posti, L. Thermal stability of a weakly magnetized rotating plasma. *MNRAS*. 2013, 428, 815.

[5] Soler, R.; Ballester, J. L.; Parenti, S. Stability of thermal modes in cool prominence plasmas. *Astronomy and Astrophysics*. 2012, 540, A7.

[6] Hobbs, A.; Read, J.; Power, C.; Cole, D. Thermal instabilities in cooling galactic coronae fuelling star formation in galactic discs. *MNRAS* 2013, 434, 1849-1868.

[7] Inoue, T.; Omukai, K. Thermal instability and multi-phase interstellar medium in the first galaxies. *The Astrophysical Journal*. 2015, 805, 73.

[8] Pines, D. Classical and quantum plasmas. *Journal of Nuclear Energy C: Plasma Physics*. 1961, 2, 5.

[9] Haas, F. A magnetohydrodynamic model for quantum plasmas. *Physics of Plasmas*. 2005, 12, 062117.

[10] Masood, W.; Sallimullah, M.; Shah, H. A. A quantum hydrodynamic model for multicomponent quantum magnetoplasma with Jeans term. *Physics Letters A*. 2008, 372, 6757-6760.

[11] Ibanez, S. H. M. Sound and thermal waves in a fluid with an arbitrary heat-loss function. *Astrophysical Journal*. 1985, 290, 33-46.

[12] Uberio, C. Electron-Inertia Effects on the transverse gravitational instability. *Journal Plasma Fusion Res. Series*. 2009, 8, 823.

[13] Bora, M. P.; Talwar, S. P. Magnetothermal instability with generalized Ohm's law. *Physics of Fluids B*. 1993, 5(3), 950.

[14] Devlen, E.; Rennan Pekunlu, E. Finite Larmor radius effects on weakly magnetized dilute plasmas. Mon. Not. R. *Astronomy Society*. 2010, 404(2), 830- 836.

[15] Sharma, S.; Sutar, D. L.; Pensia, R. K.; Patidar, A. *Effect of electron inertia on radiative instability of optically thick plasma.* AIP Conference Proceedings. 2019, 2100, 020005.

[16] Kumar, V.; Sutar, D. L.; Pensia, R. K. Effect of photoelectron current on jeans instability of rotating quantum dusty plasma. *AIP Conference Proceedings*. 2019, 2100, 020074.

ABOUT THE EDITOR

- Dr. Sachin Kaothekar is working as an Associate Professor & Head in Department of Physics, Mahakal Institute of Technology & Management, Ujjain since August 2003 and teaching Engineering students the subject Engineering Physics and Nuclear Instrumentation in their B. Tech I Year & B. Tech IV Year. He is also working as I Year in-charge in the college from 8 years. He is doing research in field of Plasma Physics since 2007 and published more than 30 papers in International Journals of repute, and 5 papers in National Journals. His area of research interest is Plasma instabilities specifically Jeans-gravitational instability, Thermal instability, Radiative instability, Dusty Plasma, Quantum Plasma and Viscoelastic fluids. He has visited abroad two times for presentation of research work in international conferences. He is member of following bodies

- **Life Membership of Plasma Science Society of India (PSSI) (LM-821).**

- **Life Membership of Association of Asia Pacific Physical Societies Division of Plasma Physics (APPC-DPP) (LM-914).**
- **Life Membership of Indian Physical Society (LM-1100).**
- **Life Membership of Astronomical Society of India (LM-2211).**
- **Life Member of Indian Society of Atomic and Molecular Physics (ISAMP) (LM-1631).**
- **Member of Institute of Economic Development and Social Research of Turkey (IKSAD).**
- **Member of International Association of Mathematical Physics.**
- **Life Member of American Association of Science and Technology (AACSIT).**
- **Senior Member International Association of Computer Science and Information Technology (IACSIT) No. - 80344907.**

ORGANIZING / TECHNICL COMMETTEE MEMBER OF CONFERENCES / WORKSHOPS / SYMPOSIA:

1. Technical Program Committee Member in, *The 3rd Conference on Astrophysics and Space Science (APSS2017)*, Bangkok, Thailand, 3-5 Jan. 2017.

2. Technical Program Committee Member in, *2017 International Conference on Engineering Physics and Optoelectronic Engineering*, Thammasat University Rangsit Campus, Bangkok, Thailand, 21-23, April, 2017.

3. Technical Program Committee Member in, *2020 International E-Conference on Plasma Theory & Simulation (PTS 2020)*, Department of Pure and Applied Physics, Guru Ghasidas Central University, Bilaspur (C.G.), India, 14-15 September 2020.

LIST OF CONFERENCE/WORKSHOP/ SYMPOSIUM CONDUCTED:

1. **2-week ISTE STTP on "Engineering Physics"**, MIT Ujjain with, IIT Bombay, 8th -18th Dec, 2015.

2. **"One Week Workshop on Nano Electro-mechanical Systems"**, Under RGPV TEQIP-3, MIT Ujjain, 18 May- 22 May 2020.

3. **"One Week Workshop on Importance of Applied Science and Humanities in Engineering"**, Under RGPV TEQIP-3, MIT Ujjain, 04 August - 08 August 2020.

INDEX

A

C

D

E

F

G

H

I

J

K

L

M

N

P

Q

R

S

T

U

V

W